Corrections
for the first printing of
Beyond Capability Confusion

In order to meet a deadline for using *Beyond Capability Confusion* in classes in Europe, some mistakes were not caught before the manuscript was sent to the press. Of course these errors, which were formerly invisible, became as obvious as a pimple on a teenager's nose once the book was printed.

The following list includes those errors that are known to date.

- Page 16, lines 5 and 6 of Section 3.3:
  Example Three became Example 4.5 in the final version.

- Page 16, line 18 of Section 3.3:
  Examples One and Two became Examples 4.1 and 4.2 in the final version.

- Page 33, line 5:
  The value 2.49 should be 2.51. This will have a small impact on the subsequent computations in this example.

- Page 35, entry for Cp = .5 and Cpk = .14 should be 34.2

- Page 37, Paragraph 3, Line 3:
  "… in the realm of 100 parts per million or less."
  should be replaced by
  "… in the realm of hundreds of parts per million or less."

  and on the next line
  "… fewer than 100 parts per million …"
  should be replaced by
  "… fewer than 1000 parts per million …"

- Page 38, line 14 from bottom:
  The word "…vary…" should be "…very…"

- Page 45, second line from bottom, replace "…a being…" with "…as being…"

- Page 57, fifth line from the bottom, "Figure 6.2" should be "Figure 6.3"

- Page 62, line 7 from the bottom:
  An extra word crept into this sentence. Delete the word "they" from this line.

- Page 84, line 2 below the table:
  "approximately" is misspelled.

- Page 85, Example 9.2: The value for K should be $ 222.2 per inch squared, rather than 2.222. This will make the ACU value 100 times larger, namely $0.0031.

- Page 99, Figure 9.18:
  Replace the word *target* with the word *nominal* in the figure.

- Page 102:
  The last two entries (9.84 and 10.84) in the first column (the column headed 1.00) are incorrect. They should be changed to 9.41 and 10.00 respectively, to match the last column of page 101.

- Page 103, line 18:
  This process will have minimum Average Cost-of-Use when B = +0.5 rather than the value of B = –0.5 shown. As a result there should be a sign change in line 19, giving a final value of 500.7 instead of 499.3.

- Page 103, line 26:
  The value of 499.7 should be 500.3.

- Page 115, line 3:
  The symbol used is wrong: "… mean ξ …" should be changed to "… mean μ …"

July 20, 1999

Beyond Capability Confusion

Donald J. Wheeler

SPC Press
Knoxville, Tennessee

Copyright © 1999 SPC Press
Statistical Process Controls, Inc.

All Rights Reserved

Do Not Reproduce
the material in this book
by any means whatsoever
without written permission
from SPC Press

SPC Press
5908 Toole Drive, Suite C
Knoxville, Tennessee 37919
(423) 584–5005
Fax (423) 588–9440

ISBN 0–945320–51-5

iv + 120 pages
75 figures
31 tables
32 examples

1 2 3 4 5 6 7 8 9 0

Contents

One	**Introduction**		1
	1.1	The Specification Game	1
	1.2	The Capability Game	2
	1.3	Predicting the Future	3
Two	**The Four Possibilities for Any Process**		5
	2.1	Two Definitions of Trouble	5
	2.2	The Four Possibilities	6
	2.3	The Effect of Entropy	9
	2.4	The Implications for Assessing Process Capability	11
Three	**Predictable and Unpredictable Processes**		13
	3.1	Two Types of Variation	13
	3.2	Two Types of Action	14
	3.3	Two Types of Process Behaviors	16
	3.4	Implications for Assessing Process Capability	17
Four	**Capability for Predictable Processes**		19
	4.1	Capability	19
	4.2	Capability Indexes	20
	4.3	A Capability Checklist	23
	4.4	Another Way to Think About the Capability Indexes	29
	4.5	Converting Capability Ratios into Fractions Nonconforming	33
	4.6	The Fraction Nonconforming When Capability Indexes Exceed 1.00	38
	4.7	Summary	40
Five	**What Can Be Said for Unpredictable Processes?**		41
	5.1	Unpredictable Processes	41
	5.2	What About Performance Indexes?	48
	5.3	Hypothetical Capability	51
	5.4	A Flowchart for Process Capability	52
	5.5	Summary	54
Six	**Capability Ratios Vary**		55
	6.1	Capability Indexes Over Time	57
	6.2	Charting Capability Indexes is No Substitute for Charting the Process	61
	6.3	So What Do Capability Indexes Do?	62

Seven	**What is Capability for Count Data?**	**63**
	7.1 Can We Calculate a Centered Capability Ratio?	64
	7.2 Capability for Count Data	64

Eight	**Why Specifications Don't Work**	**67**
	8.1 Why Specifications Don't Work	67
	8.2 Quick Fix Number One	69
	8.3 Quick Fix Number Two	70
	8.4 Quick Fix Number Three	71
	8.5 Quick Fix Number Four	72
	8.6 Quick Fix Number Five	73
	8.7 Quick Fix Number Six	74
	8.8 A New Approach to Production: Continual Improvement	75

Nine	**The Average Cost-of-Use**	**79**
	9.1 Variation Always Creates Costs	79
	9.2 The Costs of Using Conforming Product	80
	9.3 The Average Cost-of-Use	81
	9.4 The Loss Function for Conformance to Specifications	88
	9.5 A More Realistic Loss Function	90
	9.6 Approximating a More Realistic Loss Function	91
	9.7 The Average Cost-of-Use for a Predictable Process	92
	9.8 World-Class Quality	94
	9.9 Computing the Average Cost-of-Use for Your Process	97
	9.10 But What About a Loss Function That Is Not Symmetric?	98
	9.11 Summary	105

Ten	**So What Do You Do Now?**	**107**
	10.1 The Capability Game	107
	10.2 World-Class Quality	107
	10.3 "Be Ye Warmed and Filled" Is Not Enough	108

Appendix
Bibliography 113
The Origin of Table 9.2 115
Tables 116 - 119

Chapter One

Introduction

1.1 The Specification Game

In the days of mass production specifications were frequently set by the customer in the following manner: (1) guess how much variation you can tolerate for a given characteristic on a given item, (2) cut this amount of variation in half, then (3) call this reduced amount of variation the specifications.

Of course this system placed the supplier in a bind—he had to meet these specifications imposed by the customer. Yet the specifications did not tell him how to make conforming product, nor did they tell him how to avoid making nonconforming product. They simply provided a way to sort the good stuff from the bad stuff at the end of the line. Make enough stuff and some of it was bound to be good. As a result, a typical manufacturing operation became a complex cycle of fabrication, inspection, and rework.

The natural consequence of this use of specifications was added complexity for the manufacturer. While manufacturers understood that this added complexity hurt productivity, they had no way to do anything about it. It was a way of life. Industry after industry had become accustomed to the process of making, inspecting, sorting and assembling "good parts." "Bad parts" were scrapped or reworked, and those that were reworked were inspected once again. As a result, manufacturing became an attempt to meet production requirements by using the sorting process. If enough "good" units were not produced, then marginal units would be used in order to meet the shipping schedule.

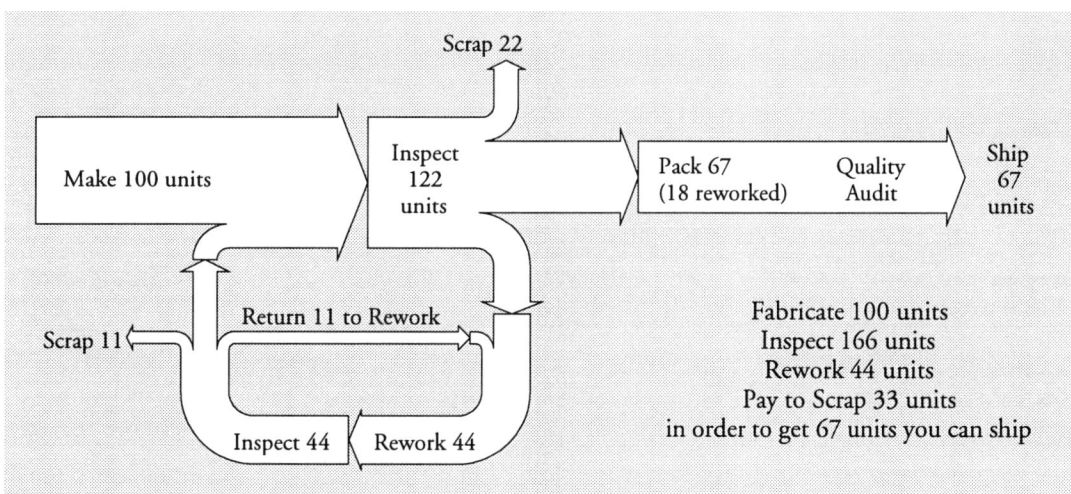

Figure 1.1: A Typical Mass Production Process

From this, the next step was the acceptance of "deviations" from the specifications. This happened because it was often necessary to use all, or nearly all, of the parts produced in order to meet the production schedule. Thus, instead of using the specifications to ensure that all parts met requirements, manufacturers tried to change the specifications in order to use as many parts as possible. This was inevitable because the producer could not use the specifications to find out why nonconforming parts were produced. He was operating in the dark. All he could do was hope for the best, and when he failed, his customer had to suffer the consequences.

Out of this conflict came a perpetual argument about how good parts had to be. Manufacturers always sought relaxed specifications. Customers demanded tighter specifications. Engineers were caught in the middle.

This conflict obscured the original and fundamental issue—how to manufacture parts with as little variation as possible. Manufacturers lost sight of the fact that if dimensions were virtually identical there would be no need to worry about "good" and "bad" pieces. There would be no need for sorting, for a scrap budget, or for rework. All the parts would fit together and work properly without the added expense and extra handling. This original objective was forgotten in the scramble to meet yesterday's production quotas with today's scrap and rework.

With the shift from mass production to lean manufacturing this burden of inspection and rework becomes intolerable. The added expense and the unpredictability of this whole approach simply have no place in a lean manufacturing environment. Therefore, instead of simply throwing the specifications to the supplier, the customer now wants to know if the supplier can meet the specifications without having to resort to inspection and rework. Now they play the Capability Game.

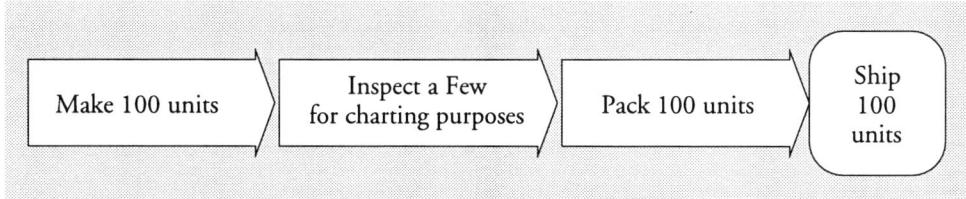

Figure 1.2: A Typical Lean Production Process

1.2 The Capability Game

Your customer wants to know if you will be able to provide him with product that conforms to the specifications without having to resort to 100 percent screening and inspection.

Your sales representative assures the customer that all of your processes are capable, and that you wouldn't think of shipping anything that wasn't conforming.

The customer, being wise to the promises of sales representatives, asks for some evidence that you can produce conforming product without resorting to sorting.

So what type of evidence can you offer?

This is the essence of the capability question.

Of course, without an agreement, in advance, about how to characterize a process there is a lot of room for the supplier to manipulate the data until it looks good. The logical consequence of this approach is the experience described by the quality director for a large company when he said: "We tried this SPC stuff but it didn't work."

"What did you try to do?" I asked.

"Well, we had all of our suppliers send us their capability numbers."

"And what was the result of doing that?"

"Oh, the numbers always looked good, but the parts were still just as bad as they had always been."

I am afraid that this experience is all too common. When the supplier is pressured to meet capability targets he quickly learns how to distort the data to make things look favorable. This has the short-term advantage of getting the customer off the supplier's back, but it has the long-term disadvantage of getting rid of the customer.

Therefore, what is needed is a way to characterize a process that is fair to both the customer and the supplier, and that at the same time is self validating, so that all parties can see that it is fair and appropriate. The purpose of this book is to describe a way to do this.

1.3 Predicting the Future

Data are historical. All data analysis is historical. All the questions of interest pertain to the future. How can you bridge this gap?

If you want to know the fraction nonconforming that you received from supplier XYZ last month, then the historical data can provide an answer. Whether your data comes from a 100% inspection, or from a sample drawn from the shipment, the historical data will allow you to obtain a number for the fraction nonconforming. No fancy computations are needed.

But if you want to know what to expect in the future, then things become a little bit more complicated. How can you use the data from the past to characterize the future? Under what conditions will this make sense? And when does such an extrapolation become more a matter of wishful thinking than a reasonable expectation?

Capability is concerned with characterizing what a process will produce in the future. You already have the accident statistics to characterize the past. Therefore we will have to begin with a discussion of *how to use the data from the past to characterize the future.*

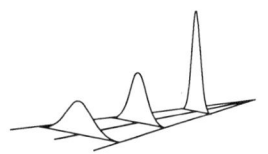

Chapter Two

The Four Possibilities for Any Process

2.1 Two Definitions of Trouble

In the past there was only one criterion required to be a good supplier: you had to ship very few nonconforming items. If your proportion of nonconforming items took a turn for the worse then you would be "in trouble," and you would stay in trouble until your fraction nonconforming dropped back down to "an acceptable level." Based on this one criterion most suppliers would alternate between being in trouble and operating okay. As long as you were operating okay you could have an attitude of benign neglect toward your operations. But when you were in trouble, then you would have to bring in the problem-solving team to fix the process. The world of the manufacturer was characterized by alternating periods of benign neglect and intense panic which could be summarized by a single dimension, as shown in Figure 2.1.

Figure 2.1: The Old Definition of Trouble

Today there is a new definition of trouble. This second definition of trouble focuses on the production process rather than the product. If your production process is predictable, then you are operating okay, but if your production process is unpredictable, then you are in trouble. The methodology for characterizing a process as being either predictable or unpredictable is the process behavior chart (also known as the control chart). Thus, the new definition of trouble would look like Figure 2.2.

Figure 2.2: The New Definition of Trouble

When these characterizations of *product* and *process* are combined we have four distinct categories:
1. Conforming and Predictable (*no trouble*),
2. Nonconforming and Predictable (*trouble, old style*),
3. Conforming yet Unpredictable (*trouble, new style*), and
4. Nonconforming and Unpredictable (*double trouble*).

2.2 The Four Possibilities

The four possibilities outlined on the previous page apply to every process.

The Ideal State (no trouble)

For lack of a better name, denote the first of these four categories as the "Ideal State." A process in this state is predictable and is producing 100 percent conforming product. The predictability of the process will be the result of deliberate efforts on the part of the manufacturer. A predictable process is an achievement, requiring constancy of purpose and the effective use of process behavior charts. The conformity of the product will be the result of having natural process limits that fall inside the specification limits.

When your process is in the Ideal State you, and your customer, can expect the conformity of the product to continue as long as the process behavior remains predictable. Since the product stream for a predictable process can be thought of as being homogeneous, the measurements taken to maintain the process behavior chart will also serve to characterize the product produced by the predictable process.

How does a process get to be in this Ideal State? Only by satisfying four conditions:
1. The process must be inherently predictable over time.
2. The manufacturer must operate the process in a predictable and consistent manner.
    The operating conditions cannot be selected or changed arbitrarily.
3. The process average must be set at the proper level.
4. The natural process limits must fall inside the specification limits for the product.

Whenever one of these conditions is not satisfied, the possibility of shipping nonconforming product exists. When a process satisfies these four conditions, the manufacturer can be confident that nothing but conforming product is being shipped. The only way that a manufacturer can know that these four conditions apply to his process, and the only way that he can both establish and maintain these conditions day after day, is by the use of process behavior charts.

For reasons that will be explained later, getting your process in the Ideal State is not the end of the continual improvement journey, but it certainly is better than the following states.

The Threshold State (trouble, old style)

Again, for lack of a better name, denote the second of these four categories as the "Threshold State." A process in this state will be predictable, but it will be producing some nonconforming product. As before, the predictability of the process will be the result of deliberate and persistent efforts on the part of the producer—a predictable process does not occur by accident. Moreover, because the process is predictable, it must be thought of as operating as consistently as it currently can operate.

Nevertheless, the existence of some nonconforming product will be the result of one or both of the natural process limits falling outside the specification limits.

The fact that the process is predictable puts a new twist on this old style of trouble. First of all, as long as the process remains predictable, the nonconforming product will persist. Therefore you cannot wait for things to spontaneously improve. Second, the ultimate solution to the problem of nonconform-

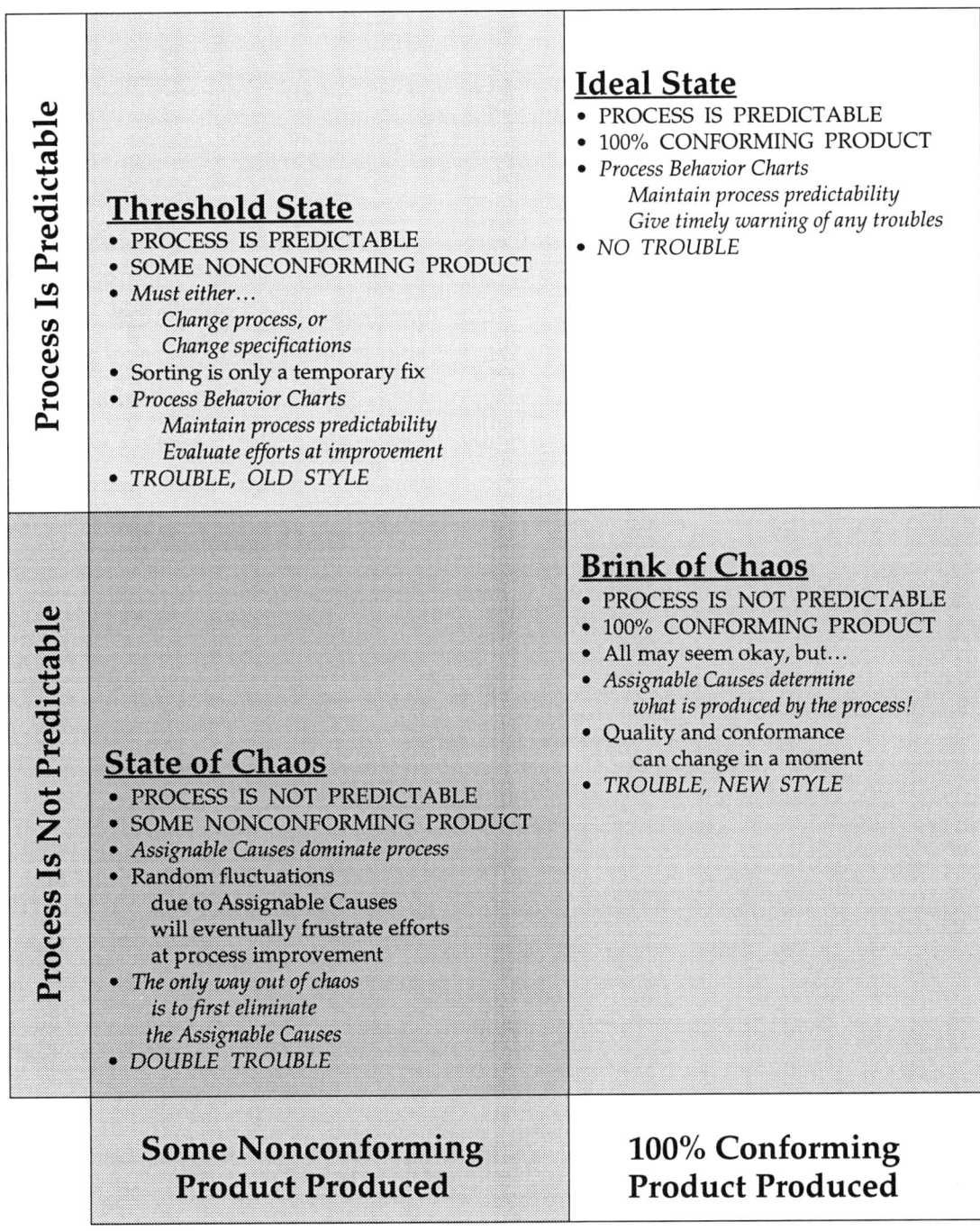

Figure 2.3: The Four Possibilities for a Process

ing product will require you to move this process up to the Ideal State. This will only happen when you either change the process or change the specifications.

If the nonconforming product occurs because the process average is not where it needs to be, then the manufacturer will need to find some way of adjusting the process aim. Here the process behavior chart can help to determine when to make adjustments and when to refrain from making adjustments to the process aim.

If the nonconforming product occurs because the natural process limits are wider than the specification limits, then you will need to try to reduce the process variation. Since a predictable process is already operating as consistently as it currently can operate, the reduction of the process variation will usually require a major change in the process itself. As you experiment with major process changes the process behavior chart will allow you to evaluate the effects of your changes. Thus, process behavior charts will not only help you achieve a predictable process, but they also help in moving the process from the Threshold State to the Ideal State.

Since making a major change in the process will often require a lot of work, you may instead try to get the specifications relaxed. With the process behavior chart to demonstrate and define the consistent *Voice of the Process*, you will at least have a chance to get the customer to agree to a change in the specifications. Without the process behavior chart to demonstrate the predictability of your process you are not likely to get the specifications changed.

As always, a short-term solution to the existence of nonconforming product is to use 100% nondestructive testing. And as has been proven over and over again, 100% screening of product is imperfect and expensive. The only way to guarantee that you will not ship any nonconforming product is to avoid making any in the first place. Sorting should be nothing more than a stop-gap measure, rather than a way of life.

Thus, process behavior charts are not only essential in getting any process into the Threshold State, but they are also critical in any attempt to move from the Threshold State to the Ideal State.

The Brink of Chaos (trouble, new style)

The third state is the "Brink of Chaos." Processes in this state are unpredictable even though they are currently producing 100% conforming product. With the traditional view the existence of 100% conforming product is considered to be evidence that the process is "operating okay." Unfortunately, this view inevitably leads to benign neglect.

The new way of characterizing process behavior emphasizes that processes on the Brink of Chaos are unpredictable—in spite of the current 100% conformity, the process is changing unpredictably, and the 100% conformity can disappear at any time.

The reason that these processes are unpredictable is that they are subject to the effects of assignable causes. These effects can best be thought of as changes in the process that apparently occur at random times. So while the conformity to specifications may lull the producer into thinking all is well, the assignable causes will continue to change the process until it will eventually produce some nonconforming product. The producer will suddenly discover that he is in trouble, yet he will have no idea of how he got there, nor any idea of how to get out of trouble. The change from 100% conforming product to some nonconforming product can come at any time, without the slightest warning. When this change occurs the process will be in the "State of Chaos."

There is no way to predict what such a process will yield tomorrow, or next week, or even in the next hour. Since the unpredictability of such processes is due to assignable causes, and since assignable causes are dominant causes that are not being controlled by the manufacturer, the only way to move out of the Brink of Chaos is to first eliminate the assignable causes. This will require the use of process behavior charts.

The State of Chaos (double trouble)

The State of Chaos exists when an unpredictable process is producing some nonconforming product.

The unpredictable process means that the producer is confronted with a changing level of nonconformity in the product stream. So even though he may know that he is making nonconforming product, he cannot reliably predict the percentage nonconforming from hour to hour.

A manufacturer whose process is in the State of Chaos knows that he has a problem, but he usually does not know what to do to correct it. Moreover, his efforts to correct the problem are ultimately frustrated by the random changes in the process which result from the presence of the assignable causes. When he makes a needed modification to the process, the effect will be short-lived because the assignable causes continue to change the process. When he makes an unnecessary modification, a fortuitous shift by the assignable causes may mislead him. No matter what he tries, nothing works for long because the process is always changing. As a result, he finally despairs of ever operating his process rationally, and begins to speak in terms of "magic" and "art."

The only way to make any progress in moving a process out of the State of Chaos is to first eliminate the assignable causes. This will require the disciplined and effective use of process behavior charts. As long as assignable causes are present, the manufacturer will find his efforts to be like walking in quicksand. The harder he tries to get free, the more deeply mired he becomes.

2.3 The Effect of Entropy *(the cause of much trouble)*

All processes belong to one of these four states. But processes do not always remain in one state. It is possible for a process to move from one state to another. In fact there is a universal force acting on every process that will cause it to move in a certain direction. That force is entropy. It continually acts upon all processes to cause deterioration and decay, wear and tear, breakdowns and failures.

Entropy is relentless. Because of it every process will naturally and inevitably migrate toward the State of Chaos. The only way this migration can be overcome is by continually repairing the effects of entropy. Of course this means that the effects for a given process must be known before they can be repaired. With such knowledge, the repairs are generally fairly easy to make.

On the other hand, it is very difficult to repair something when one is unaware of it. Yet if the effects of entropy are not repaired, it will come to dominate the process, and force it inexorably toward the State of Chaos.

The Cycle of Despair (the result of the old definition of trouble)

Since everybody knows that they are in trouble when their process is in the State of Chaos it is inevitable that problem-solvers will be appointed to drag the process up from the State of Chaos. With luck, these problem-solvers can get the process back to the Brink of Chaos—a state which is erroneously considered to be "out-of-trouble" in most operations.

Once they get the process back to the Brink of Chaos the problem solvers are sent off to work on another problem. As soon as their backs are turned, the process begins to move back down the entropy slide toward the State of Chaos.

New technologies, process upgrades, and all the other "magic bullets" which may be tried can never

Beyond Capability Confusion

overcome this Cycle of Despair. One may change technologies—often a case of jumping out of the frying pan and into the fire—but the benign neglect which inevitably occurs when the process is on the Brink of Chaos will allow entropy to drag the process back down to the State of Chaos. Thus, focusing solely upon conformance to specifications will condemn you to forever cycle between the State of Chaos and the Brink of Chaos.

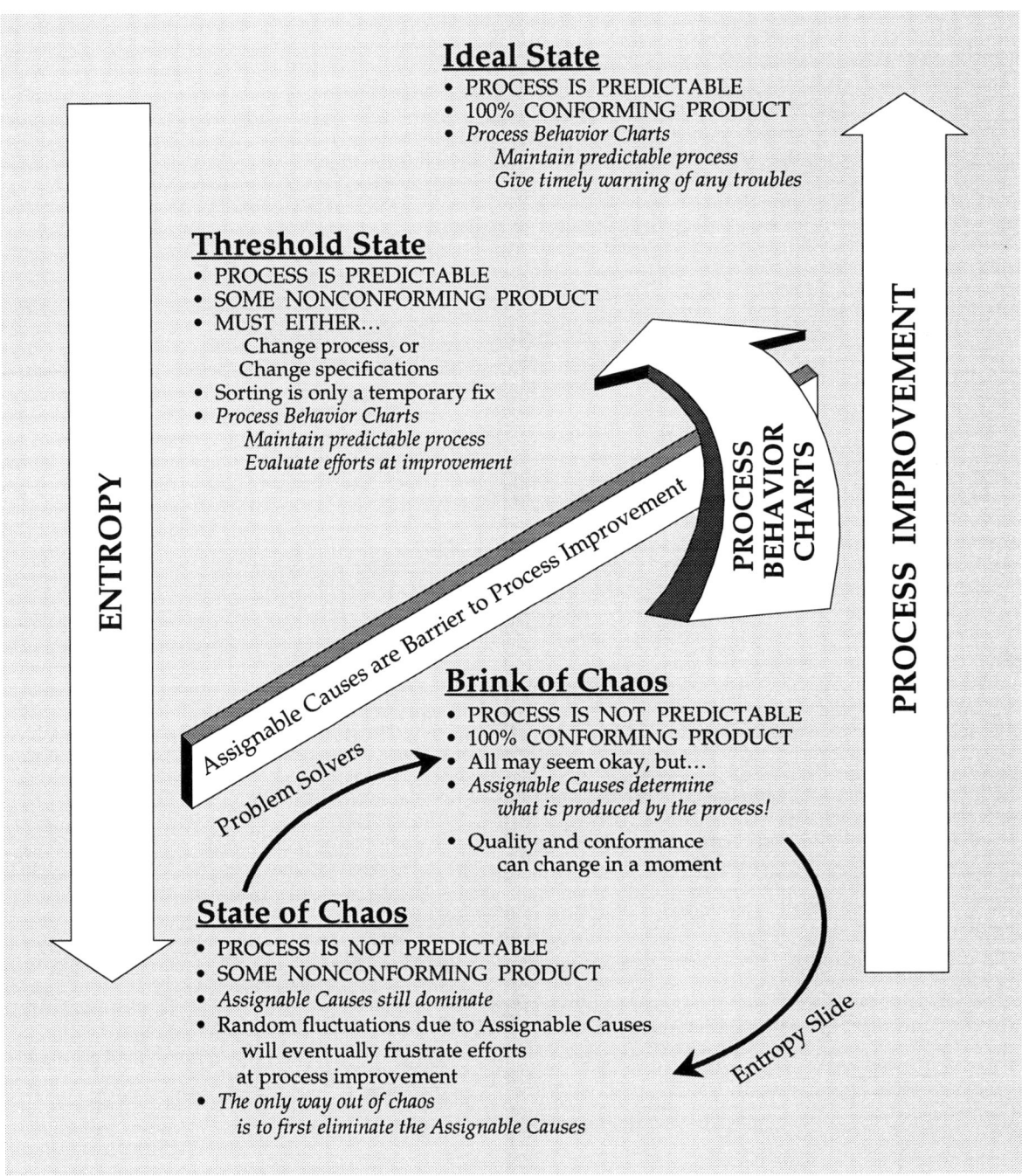

Figure 2.4: The Only Way Out of the Cycle of Despair

The Only Way Out

There is only one way out of this Cycle of Despair. There is only one way to move a process up to the Threshold State or the Ideal State—and that is to use process behavior charts effectively for process improvement.

Every manufacturer is confronted with a dual problem. Entropy places a process in the Cycle of Despair. Assignable causes doom it to stay there. Thus, manufacturers must identify both the effects of entropy and the presence of assignable causes. The only way to do this is to use process behavior charts. No other tool will consistently and reliably provide the necessary information in a clear and understandable form.

The traditional chaos-manager, trouble-shooting approach is focused upon conformance to specifications. It does not attempt to characterize or understand the behavior of the process. Therefore, about the best that it can achieve is to get the process to operate in the Brink of Chaos some of the time.

> ...which is why any process
> operated without the benefit of process behavior charts
> is ultimately doomed to operate in the State of Chaos...

2.4 The Implications for Assessing Process Capability

While conformance to specifications is important, the fundamental concept that some processes are predictable, while others are not, makes the issue of conformity to specifications an issue *which cannot be addressed directly.* If a process is predictable, then its conformity or nonconformity will also be predictable. If a process is unpredictable, then its conformity will be unpredictable, and anything we say about the *process* will amount to little more than wishes and hopes. With an unpredictable process we may attempt to characterize the extent of today's conformity, but such a characterization will not tell us anything about tomorrow's conformity.

Thus, the four possibilities for any process are unavoidable. While they complicate the job of assessing process capability, we cannot ignore them. Since the four possibilities reflect reality, any attempt to assess *process* capability that does not take these different *process* conditions into account will inevitably be flawed. Therefore, we will have to look at capability for different situations, rather than using a single approach for all conditions.

Finally, in order to be fair to both supplier and customer, an operational definition of a capable process will be needed. This operational definition will be given following the discussion of the issues involved in process capability.

But first, the notion of predictable and unpredictable processes must be considered more carefully.

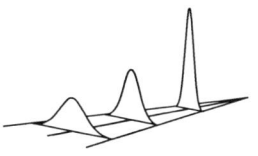

Chapter Three

Predictable and Unpredictable Processes

What is the difference between a predictable process and an unpredictable process and how can you spot this difference? The first step to answering these questions is to make a distinction between two types of variation.

3.1 Two Types of Variation

No matter what your process, no matter what your data, all data display variation. Any measure you can think of that will be of interest to your business will vary over time. The reasons for this variation are many. There are all sorts of causes that have an impact on your process and the process outcomes. It is not unrealistic to think that your processes and systems will be subject to dozens, or even hundreds, of cause-and-effect relationships. And this multiplicity of causes has two consequences: it makes it easy for you to pick out an explanation for why the current value is so high, or so low; and it makes it very hard for you to know if your explanation is even close to being right.

So in your heart, you know that there is considerable uncertainty in your interpretation of the current value of any measure. But what are you going to do about this? How can you interpret the current value when the previous values are so variable? The key to understanding any time series is to make a distinction between two types of variation.

The first type of variation is routine variation. It is always present. It is unavoidable. It is inherent in the process. Because this type of variation is routine, it is also predictable. The second type of variation is exceptional variation. It is not always present. It is not routine. It comes and it goes. Because this type of variation is exceptional, it will be unpredictable.

The first benefit of this distinction is that it provides a way to know what to expect in the future, which is the essence of management.

While every process displays variation,
some processes display predictable variation,
while others display unpredictable variation.

- A process that displays predictable variation is consistent over time. Because of this consistency we may use the past as a guide to the future.

- A process that displays unpredictable variation is changing over time. Because of these changes we cannot depend upon the past to be a reliable guide to the future.

Having made this *conceptual* distinction between predictable processes and unpredictable processes, we now need a way to do the same in *practice*. We begin by observing that while a predictable process will be the result of routine variation, an unpredictable process will possess both routine variation and exceptional variation. Thus, if we can devise a way to detect the presence of exceptional variation, we will be able to characterize our processes as being predictable or unpredictable.

Beyond Capability Confusion

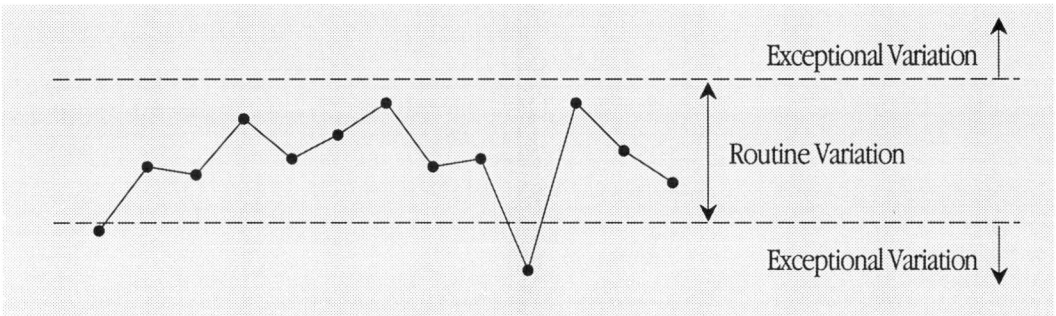

Figure 3.1: Separating Routine Variation from Exceptional Variation

In order to obtain signals of exceptional variation we will compute limits for the running record of our data. As shown in Figure 3.1, the idea is to establish limits that will allow us to distinguish between routine variation and exceptional variation.

If we compute values that place the limits too close together we will get *false alarms* (or false signals) when routine variation causes a point to fall outside the lines by chance. This is the first type of mistake we could make. We can avoid this mistake entirely by computing limits that are far apart.

But if we have the limits too far apart we will *miss some signals* of exceptional variation. This is the second type of mistake we could make. We can minimize the occurrence of this mistake only by having the limits close together.

The trick is to strike a balance between the consequences of these two mistakes, and this is exactly what Walter Shewhart did when he created the *control chart*.* In his book, *Economic Control of Quality of Manufactured Product,* Shewhart found that "three-sigma" limits would effectively balance the economic consequences of getting false alarms and with those of missing some signals. Three-sigma limits help you to separate the exceptional variation from the routine variation by bracketing approximately 99% to 100% of the routine variation. As a result, whenever you have a value outside these limits you can be reasonably sure that the value is the result of exceptional variation. The use of three-sigma limits, computed in the proper way, is the essence of the process behavior chart approach to data analysis.

3.2 Two Types of Action

So what does it mean when your current value is outside the limits and is the result of exceptional variation? How can you make use of this information? To answer this we return to the fact that every process is subject to dozens of cause-and-effect relationships.

When a process displays predictable variation, that variation may be thought of as the result of many different cause-and-effect relationships *where no one cause is dominant over the others.* While every process is subject to many different cause-and-effect relationships, predictable processes are those where the net effect of the multiple causes is a sort of static equilibrium, resulting in routine variation. We could call these causes of routine variation common causes.

On the other hand, when a process displays unpredictable variation, that variation must be thought of as consisting of the sum of the routine variation plus some *additional* cause-and-effect relationships.

---

* Shewhart called routine variation "controlled variation," and exceptional variation was "uncontrolled variation." Out of this terminology Shewhart's charts came to be called control charts. However, because of the baggage associated with the word "control" this name can be quite misleading. Therefore we shall use the more descriptive name of process behavior charts to describe this versatile technique.

Since the sum is unpredictable, we must conclude that the additional cause-and-effect relationships dominate the routine variation, and therefore that it will be worth our while to identify these additional cause-and-effect relationships. For this reason Shewhart called these dominant causes assignable causes.

If we characterize the routine variation of a predictable process as noise, then the exceptional variation of an unpredictable process would be like signals buried within the noise, and a process behavior chart allows us to detect these signals because it filters out the noise.

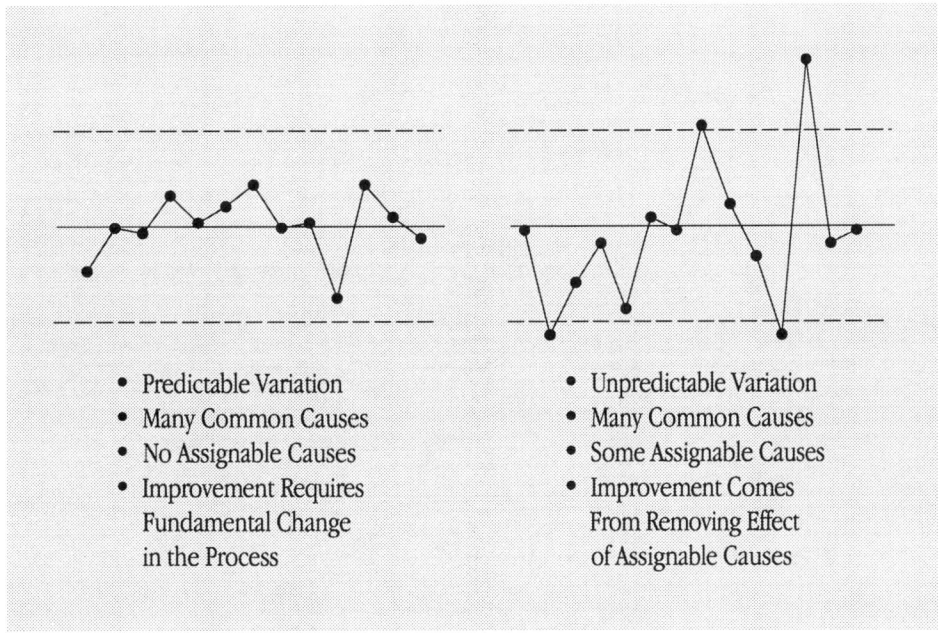

Figure 3.2: Two Types of Processes

This is how the characterization of a process as predictable or unpredictable can be the key to improving that process. If a process is predictable no one cause-and-effect relationship will stand out as the place to start. But if a process is unpredictable, the process behavior chart will allow you to detect the presence of assignable causes, so you can improve the process by working to remove the impact of these assignable causes.

Sometimes these assignable causes are external to the process—upsets due to circumstances beyond our control. At other times these assignable causes are internal to the system—former common causes that have, for whatever reason, become dominant over the other common causes.

When we identify and remove an assignable cause that was external to the process, we are, in most cases, merely returning the process to its previous condition. However, when we identify and remove the effect of an assignable cause that is internal to the process we will be improving both the process and the process outcomes. It is in the latter case that process improvement charts become the locomotive of Continual Improvement.

Thus, the path to process improvement depends upon what type of variation is present.

- If a process displays predictable variation, then the variation is the result of many common causes and it will be a waste of time to look for assignable causes. Improvement will only come by changing a major portion of the process.

- If a process displays unpredictable variation, then in addition to the common cause variation there is an extra amount of variation that is the result of one or more assignable causes. Improvement will come by finding and removing the assignable causes. Changing a major portion of the process will be premature.

Thus, the distinction between the two types of variation leads to two different routes to process improvement. When a process is unpredictable, looking for assignable causes will be a high payback strategy because the assignable causes are, by definition, dominant. On the other hand, looking for common causes of routine variation will be a low payback strategy because no single common cause is dominant. And a process behavior chart is the operational definition of an assignable cause. No guesswork is needed. Simply plot your data, compute the limits, and start to interpret the chart. Whenever points fall outside the limits, go look for the assignable causes. As you find the assignable causes, and as you take appropriate actions to remove the effects of these assignable causes from your process, you will be improving your process.

3.3 Two Types of Process Behavior Charts

The simplest type of process behavior chart is the *chart for individual values and a moving range* (also known as an *XmR* chart). The limits on the *X* portion of this chart are called Natural Process Limits. These Natural Process Limits define the range of routine variation for the process outcomes. Since the questions of capability are concerned with this range of process outcomes, we will make use of these Natural Process Limits in the following chapter. An example of an *XmR* chart may be seen in Example Three. The *XmR* chart may be used with any type of data—counts, measurements, ratios, or percentages—as long as the values are logically comparable and as long as the moving ranges reasonably represent the routine variation of the measure being plotted.

The other common type of process behavior chart is the *average and range chart* (also known as an X-bar and *R* chart). This chart will have the individual values arranged in small subgroups where each subgroup consists of values that were collected under essentially the same conditions. This chart plots the averages and ranges for these subgroups and computes limits for these summary statistics. Since averages cannot vary as much as individual values vary, *the limits for the subgroup averages may not be used directly to characterize capability*. Therefore, when an average and range chart is used, the Natural Process Limits will have to be specifically computed using the formula:

$$\text{Natural Process Limits} = \text{Grand Average} \pm 3 \, \frac{\text{Average Range}}{d_2}$$

where the value for d_2 depends upon the subgroup size. Examples of this type of process behavior chart may be found in Examples One and Two.

While there are other types of process behavior charts these two basic charts will cover virtually all applications. A complete explanation of process behavior charts is beyond the scope of this book. You may find out more about this powerful technique of data analysis in the books listed in the bibliography.

3.4 Implications for Assessing Process Capability

Before you can determine if a process is predictable you will have to have a process behavior chart. If you do not have a process behavior chart, then any assumption that you have a predictable process is almost certainly incorrect. A predictable process is not a natural state for a production process. It is instead an achievement, requiring determined and persistent effort. Which is why, in the absence of a process behavior chart that displays a reasonable degree of predictability, you should always assume that your process is unpredictable.

Given that you have a predictable process, your process behavior chart will provide the necessary information for computing the capability indexes. No special data collection is necessary. The capability of a predictable process is the same over the long term as it is over the short term. Process capability cannot be separated from process behavior, and the fact that some processes are predictable while others are unpredictable makes the whole question of process capability one that cannot be addressed directly.

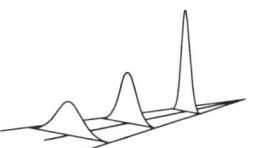

Chapter Four

Capability for Predictable Processes

Now what do we mean by the words *process capability*? Capability generally refers to that which can be achieved. It combines the connotations of potential and performance. Therefore, it is reasonable to think of the term process capability as defining *that range of product values that we can expect to get from a production process over an extended period of time.*

Based on this concept of process capability, we would then describe a process as being *capable* if this range of product values falls entirely inside the specification limits.

Since the concept of a capable process involves the comparison of a range of product values with a set of specification limits, it was natural to come up with numerical summaries that express the degree of overlap between these two ranges of values. These numerical summaries are called *capability ratios*, or *capability indexes*. Like all index numbers, capability indexes use the value of 1.00 as the borderline between favorable from unfavorable situations. Moreover, capability indexes create an artificial concept of distance which we can use in our discussions of capability—they allow us to characterize processes with regard to how close they are to the threshold of "being capable."

Thus, the basic concept is process capability: "what will a given process produce?" If this range of product values is entirely within the specifications, then the process is said to be capable, and capability indexes are merely numerical summaries to describe how close to the borderline a given process may be.

4.1 Capability

When we have a predictable process the *process capability* will be defined by the Natural Process Limits—the three-sigma limits for individual values (calculated in a specific way). When a process is predictable the Natural Process Limits will bracket virtually all of the process outcomes. Moreover, as long as the process continues to be operated predictably, the Natural Process Limits will continue to characterize the process outcomes.

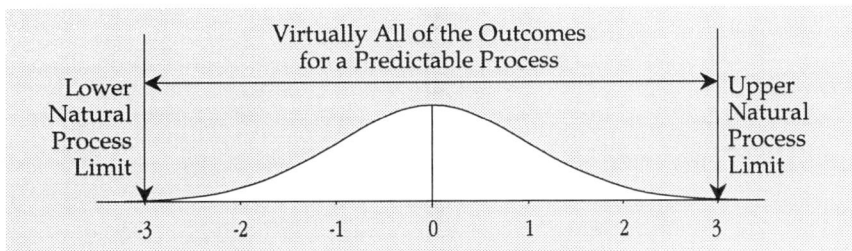

Figure 4.1: The Capability of a Predictable Process

Thus a process behavior chart will allow both the supplier and the customer to assess the predictability of the production process, and when the process is predictable the process behavior chart will also provide the information needed to characterize the process as capable or not.

A process is said to be *capable* when the range of values defined by the process capability falls within

19

the specification limits. If a predictable process has Natural Process Limits that fall inside the specification limits, then the process may be said to capable (see Figure 4.2). If one or more of the Natural Process Limits fall outside the specification limits then the process may be not capable (see Figure 4.3).

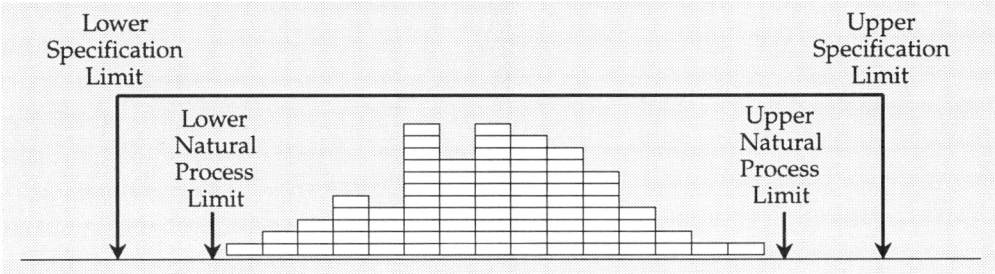

Figure 4.2: A Predictable Process that is Capable

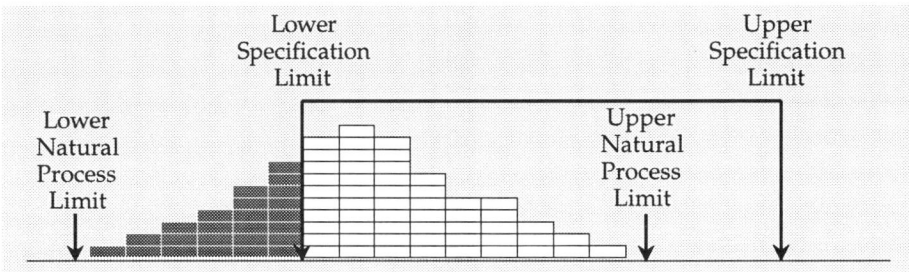

Figure 4.3: A Predictable Process that is Not Capable

Thus, for a predictable process, you can determine if you have a capable process by simply comparing the Natural Process Limits with the specification limits. No calculations beyond those required to obtain the Natural Process Limits are needed.

Figures 4.2 and 4.3 also suggest that, in addition to the process behavior chart, a useful visual summary of the process outcomes may be obtained from a histogram of the raw data from your chart. With the addition of lines to represent the specification limits this histogram becomes a powerful visual summary of the capability of a predictable process.

While such a visual summary is best, there is a tendency to try to quantify the relationship between the specification limits and the Natural Process Limits, and this is where capability indexes enter the picture.

4.2 Capability Indexes

Capability indexes have been defined in many different ways. There are dozens of formulas that have been created. Yet in the end, they are primarily concerned with only two characteristics of the process—location and dispersion. This book cannot begin to list all of the different formulas that have been created. However, since there are only two basic concepts, we will begin with these concepts and then develop the two most commonly used formulas.

As was noted earlier, whether or not a predictable process is capable depends upon how the Natural Process Limits are related to the specification limits. However, since these Natural Process Limits depend upon the process location and process dispersion, we can also characterize whether or not a predictable process is capable using another relationship—the relationship between the specification limits and the process location and dispersion. This relationship between the process characteristics and the specification limits can be summarized by two questions. In particular, is there enough room within the specification limits for the process to operate, and is the process properly located to take advantage of what room there is within the specification limits?

The first of these questions concerns what I call the "Elbow Room" for the process. This Elbow Room is defined by the distance between the specification limits, and is commonly called the specified tolerance.

Elbow Room = Specified Tolerance = Upper Specification Limit − Lower Specification Limit

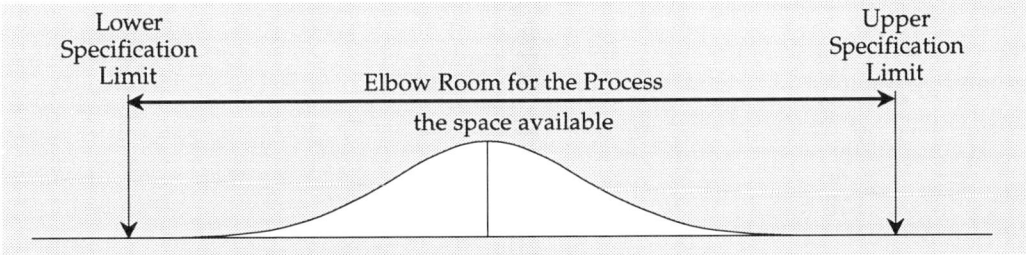

Figure 4.4: The Specified Tolerance as the Elbow Room for a Process

In its raw form the Elbow Room for a process is expressed in measurement units. It is an expression of the amount of *space available*. In order to obtain an answer to the question "Is there enough room within the specification limits for the process to operate?" the space available will have to be compared to the *space required* for the process. Since the Natural Process Limits will bracket virtually all of the outcomes from a predictable process, the space required for such a process will be the distance between the Natural Process Limits. So we compare the space available to the space required by dividing the Elbow Room by the distance between the Natual Process Limits to get the Capability Ratio:

$$\text{Capability Ratio} = C_p = \frac{\text{Specified Tolerance in measurement units}}{\text{Distance between Natural Process Limits}}$$

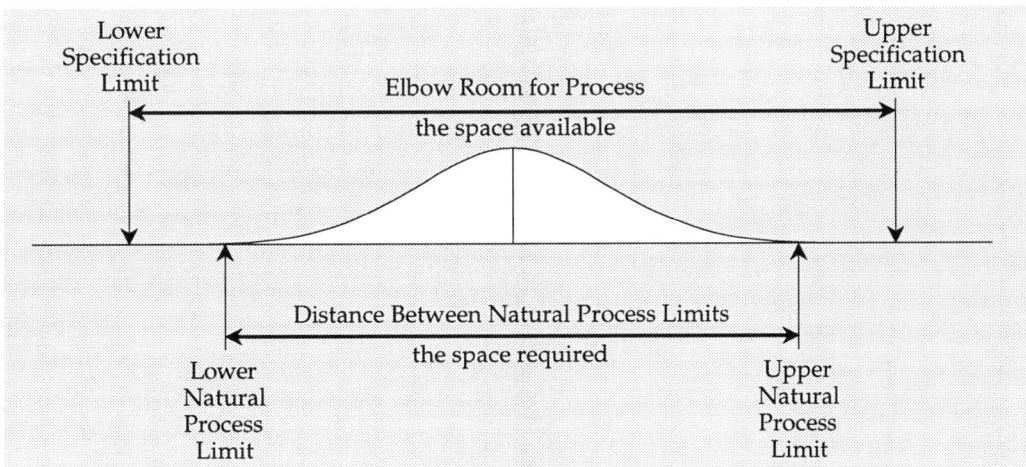

Figure 4.5: The Quantities Compared by the Capability Ratio

If the process is perfectly centered within the specification limits, then the Capability Ratio computed above will correctly describe the situation. But when the process is off-center the effective space available will be reduced. So we need to consider the location of the process relative to the specification limits. This location may be characterized by the distance between the process average and the nearer of the two specification limits:

Distance to Nearer Spec = Process Average − Nearer Specification Limit

When the process average is on the correct side of the specification limits we will define the Distance to Nearer Spec to be a positive value. If the process average should fall below a lower specification limit, or if it should happen to fall above an upper specification limit, then the Distance to Nearer Spec will have a minus sign attached as a way of signifying the undesirable location of the process.

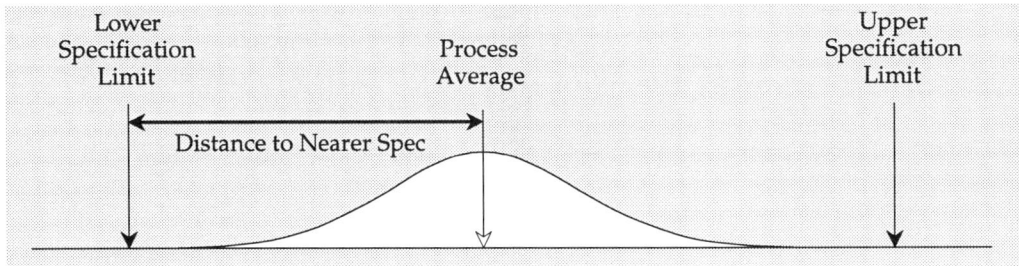

Figure 4.6: Distance to Nearer Spec as a Characterization of Process Location

In its raw form, the Distance to Nearer Spec is expressed in measurement units. It is a characterization of the *effective space available* between the process average and the nearer specification limit. In order to obtain an answer to the question "Is the process properly located to take advantage of what room exists within the specifications?", we will need to compare the effective space available with the *space required* for one side of the process. A generic value for this space required will be half of the distance between the Natural Process Limits. Thus, the comparison is made by dividing the Distance to Nearer Spec by a value that is one-half of the distance between the Natural Process Limits. This ratio is the usual form for the Centered Capability Ratio:

$$\text{Centered Capability Ratio} = C_{pk} = \frac{\text{Distance to Nearer Spec in measurement units}}{\text{One-half of distance between Natural Process Limits}}$$

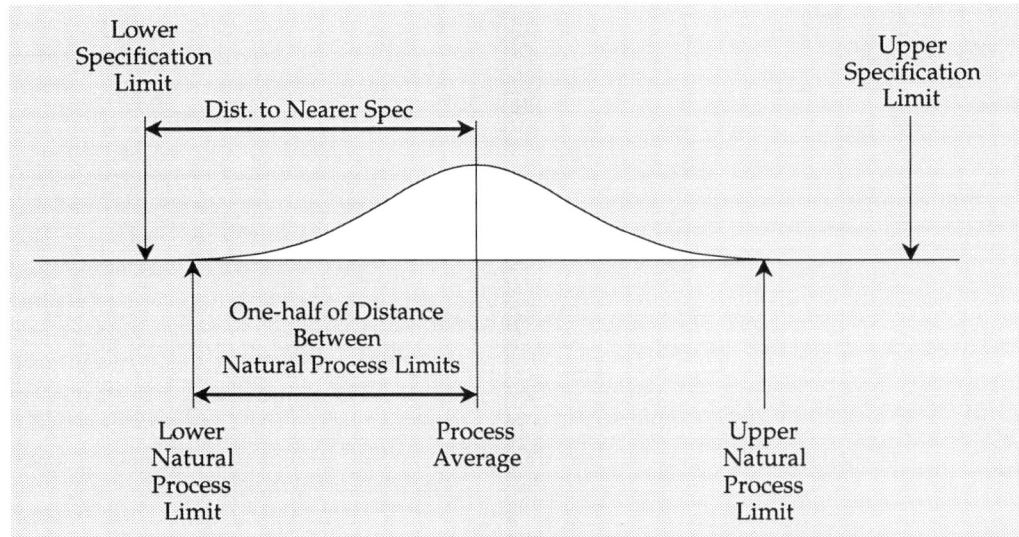

Figure 4.7: The Quantities Compared by the Centered Capability Ratio

Both the Capability Ratio and the Centered Capability Ratio are index numbers; that is, they are dimensionless numbers where values greater than 1.00 are favorable. Each of these indexes express the space available as a proportion of the space required.

When you have two specification limits you can compute both the capability indexes defined above, however, the Centered Capability Ratio cannot be greater than the Capability Ratio. For this reason the Centered Capability Ratio is the more commonly used of the two values. When you have a one-sided specification you will only be able to compute the Centered Capability Ratio.

4.3 A Capability Checklist

1. Place data from the product stream on a process behavior chart. If, over a reasonable period of time, the chart shows no evidence of unpredictable behavior, then the process may be judged to be predictable.
2. For a predictable process, create a histogram of the individual values and show the specification limits on this histogram. A simple estimate of the fraction of nonconforming product can be obtained from this histogram.
3. For a predictable process, compute the Natural Process Limits from the information provided by the process behavior chart.
4. For a predictable process, compute the Capability Ratio and the Centered Capability Ratio as appropriate.
5. Combine the process behavior chart, the histogram of individual values, and the computed capability indexes, and present all of these together as your summary of the process capability.

Example 4.1:   The Rheostat Knob Data:

1. **Place data from the product stream on a process behavior chart.**
   A particular rheostat knob had a pin hole on the shaft housing. The dimensions recorded are the distances from the back of the shaft to the near side of the pin hole. These measurements were made to the nearest thousandth of an inch. The Average and Range chart for these data reveals a process that is reasonably predictable.

Subgroup						$\bar{X}$	R	Subgroup						$\bar{X}$	R
1	.140	.143	.137	.134	.135	*.1378*	*.009*	15	.144	.142	.143	.135	.144	*.1416*	*.009*
2	.138	.143	.143	.145	.146	*.1430*	*.008*	16	.133	.132	.144	.145	.141	*.1390*	*.013*
3	.139	.133	.147	.148	.149	*.1432*	*.016*	17	.137	.137	.142	.143	.141	*.1400*	*.006*
4	.143	.141	.137	.138	.140	*.1398*	*.006*	18	.137	.142	.142	.145	.143	*.1418*	*.008*
5	.142	.142	.145	.135	.136	*.1400*	*.010*	19	.142	.142	.143	.140	.135	*.1404*	*.008*
6	.136	.144	.143	.136	.137	*.1392*	*.008*	20	.136	.142	.140	.139	.137	*.1388*	*.006*
7	.142	.147	.137	.142	.138	*.1412*	*.010*	21	.142	.144	.140	.138	.143	*.1414*	*.006*
8	.143	.137	.145	.137	.138	*.1400*	*.008*	22	.139	.146	.143	.140	.139	*.1414*	*.007*
9	.141	.142	.147	.140	.140	*.1420*	*.007*	23	.140	.145	.142	.139	.137	*.1406*	*.008*
10	.142	.137	.134	.140	.132	*.1370*	*.010*	24	.134	.147	.143	.141	.142	*.1414*	*.013*
11	.137	.147	.142	.137	.135	*.1396*	*.012*	25	.138	.145	.141	.137	.141	*.1404*	*.008*
12	.137	.146	.142	.142	.146	*.1426*	*.009*	26	.140	.145	.143	.144	.138	*.1420*	*.007*
13	.142	.142	.139	.141	.142	*.1412*	*.003*	27	.145	.145	.137	.138	.140	*.1410*	*.008*
14	.137	.145	.144	.137	.140	*.1406*	*.008*								

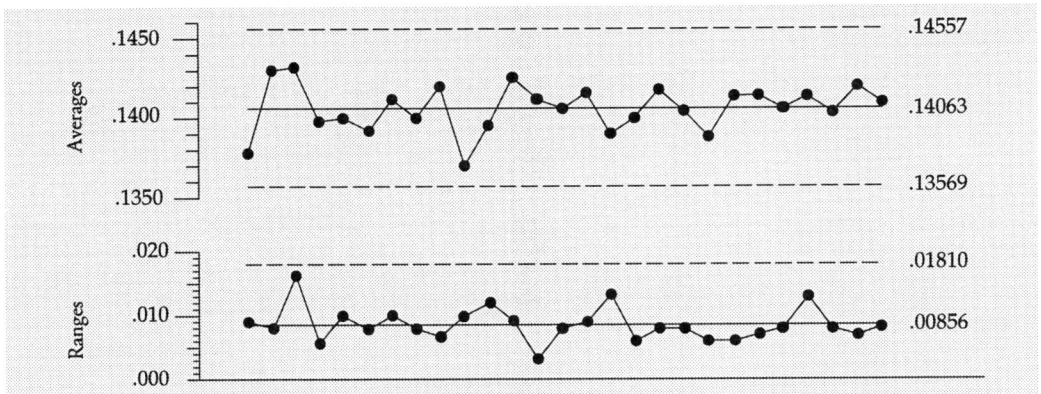

Figure 4.8: Average and Range Chart for Rheostat Knob Data

2. **For a predictable process, create a histogram of the individual values and show the specification limits on this histogram.**
   The specification limits for this dimension on this part are 0.125 to 0.155 inches. The 135 individual values shown above yield the histogram in Figure 15. Out of these 135 values none are even close to going outside the specification limits. Thus, this histogram suggests that this predictable process is capable of meeting these specifications.

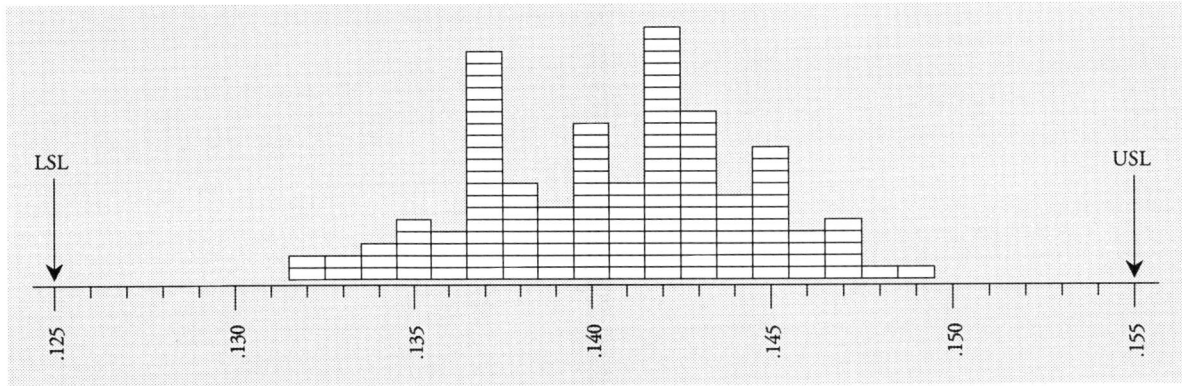

Figure 4.9: Histogram for the Rheostat Knob Data

3. **For a predictable process, compute the Natural Process Limits.**
   The process behavior chart shows the Grand Average to be 0.14063, and the Average Range to be 0.00856. For subgroups of size 5 the value for d_2 is 2.326. Thus the Natural Process Limits for individual values are:
   $$0.14063 \pm 3 \frac{0.00856}{2.326} = 0.1296 \text{ to } 0.1517$$

   These limits suggest that, as long as this process continues to be operated in a predictable manner, virtually all of the product produced by this process will fall within the interval of 0.130 to 0.152 inches.

4. **For a predictable process, compute the Capability Ratio and the Centered Capability Ratio.**

The Elbow Room for this dimension is 0.155 − 0.125 = 0.030 inches. The distance between the Natural Process Limits is 0.1517 − 0.1296 = 0.0221 inches. Thus the Capability Ratio is:

$$C_p = \frac{0.030}{0.0221} = 1.36$$

So the space available is 36 percent wider than the space required by this process.

The Distance to Nearer Spec for this process is 0.155 − 0.1406 = 0.0144 inches. Half of the distance between the Natural Process Limits is 0.01105 inches. Thus the Centered Capability Ratio is:

$$C_{pk} = \frac{0.0144}{0.01105} = 1.30$$

This process is slightly off-center, making the effective space available only 30 percent wider than the space required by this process.

5. **Summarize the process capability.**
   - The process behavior chart shows the process to be predictable.
   - The histogram shows the process to be comfortably capable.
   - The Capability Ratio of 1.36, and the Centered Capability Ratio of 1.30 both support the impression given by the histogram.
   - This process is clearly in the Ideal State, it is predictable and capable.

A predictable process with capability indexes that are greater than 1.00 will generally have a product stream with a nonconformity level that is less than one part per hundred. So while it is good to have a Capability Ratio of 1.0, it is better to have one that is slightly greater than 1.0. The Natural Process Limits are characterizations of what the *process* will produce. The specifications define that *product* which is considered to be minimally acceptable. We would like there to be a cushion between that which we produce and that which is minimally acceptable.

A predictable process with capability indexes greater than 1.3 will generally have a product stream with a nonconformity level that is less than one part per ten thousand. Which is why customers often ask for capability indexes of 1.33 or greater.

Even though the capability indexes provide a neat numerical summary they should always be used in conjunction with the process behavior chart and the histogram of individual values. Just why this is so will be seen in the next example.

Beyond Capability Confusion

Example 4.2:   The Ball-Joint Socket Thickness Data

1. **Place data from the product stream on a process behavior chart.**
   The following Average and Range chart contains one week's worth of data for an injection molded part. A subgroup consists of the four parts produced in a single shot of the press. This subgrouping works because they have managed to make the four cavities alike with regard to the thickness of the parts. The Average and Range chart shows a process that is behaving predictably. Moreover, this process has been operated predictably for the past six months.

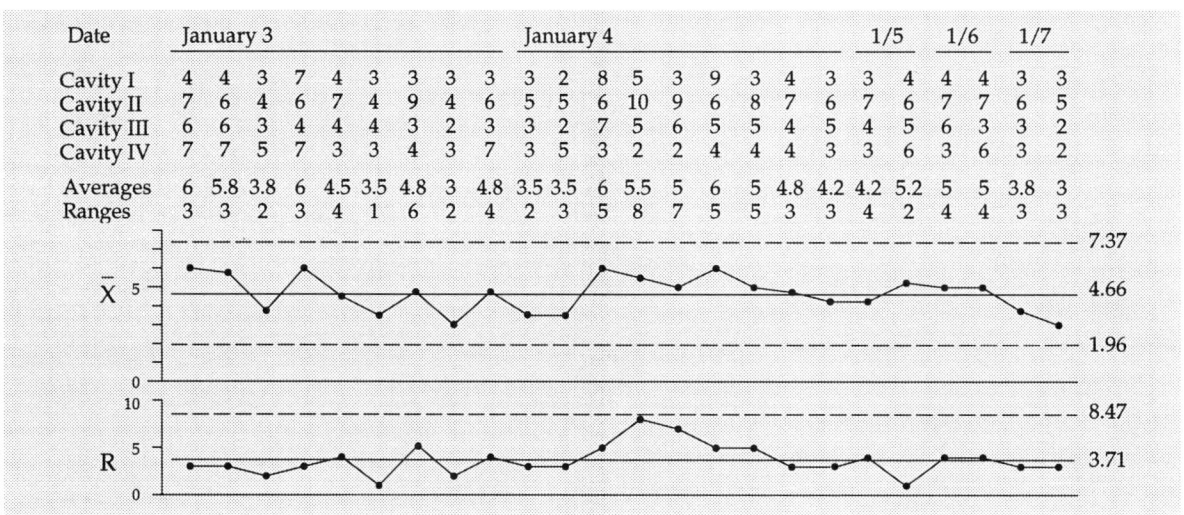

Figure 4.10: Average and Range Chart for Ball-Joint Socket Thicknesses

2. **For a predictable process, create a histogram of the individual values and show the specification limits on this histogram.**
   The process histogram in Figure 4.11 uses the 96 individual values shown on the Average and Range chart. The specification limits are 0 to 15, inclusive. As may be seen from this histogram, the estimated fraction nonconforming for this process is:

   $$\frac{0}{96} = 0.00 \quad \text{or} \quad \text{zero parts per hundred nonconforming.}$$

   Therefore, this predictable process would seem to be capable. In fact this process has been predictable and capable for the past six months.

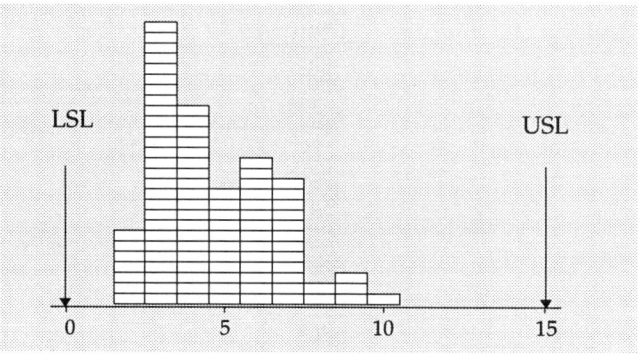

Figure 4.11: Histogram for Ball-Joint Socket Thicknesses

3. **For a predictable process, compute the Natural Process Limits.**
   The process average is approximated by the Grand Average of 4.66 units. The process dispersion is characterized by the Average Range of 3.71 units. With subgroups of size $n = 4$, the appropriate value for d_2 is 2.059. Using the formula given earlier, the Natural Process Limits for the individual values produced by this process are:

   $$4.66 \text{ units} \pm 3 \, \frac{3.71 \text{ units}}{2.059} = 4.66 \pm 5.40 = -0.74 \text{ units to } 10.06 \text{ units}$$

   Since the value of 0 corresponds to 12.75 mm, negative values are possible and do make sense for this dimension. Thus, we would expect this process to produce items whose thicknesses will vary between −1 and 10 units (12.74 mm to 12.85 mm), and we would also expect virtually all of the process output will fall within this range as long as the process is operated predictably.

4. **For a predictable process, compute the Capability Ratio and the Centered Capability Ratio.**
   The Capability Ratio for this process is the ratio of the Elbow Room to the distance between the Natural Process Limits. The Elbow Room is $15 - 0 = 15$ units. The distance between the Natural Process Limits is $10.06 - (-0.74) = 10.8$ units. The Capability Ratio is thus:

   $$\text{Capability Ratio} = \frac{15}{10.8} = 1.39$$

   This value is interpreted to mean that the space available is about 39% greater than the space required. Thus, there is elbow room for this process to operate within these specifications.

   The Centered Capability Ratio depends upon the Distance to Nearer Spec, which is:

   $$4.66 - 0 = 4.66 \text{ units.}$$

   Half the distance between the Natural Process Limits is:

   $$4.66 - (-0.74) = 5.4 \text{ units.}$$

   The Centered Capability Ratio is thus:

   $$\text{Centered Capability Ratio} = \frac{4.66}{5.4} = 0.86$$

   This value takes into account the fact the process is not centered within the specifications. Because of the location of this process, the effective space available is only 86% of the generic space required for one side of the process.

5. **Summarize the process capability.**
   - The process behavior chart shows the process to be predictable. This has been the case for the past six months.
   - The histogram shows a process that is producing 100% conforming product. This has also been the case for the past six months.
   - The Capability Ratio of 1.39 suggests that there is plenty of elbow room within the specifications.
   - The Centered Capability Ratio of 0.86 suggests that there could be some nonconforming product!
   - So what are we to believe? Do we condemn this process because it has a Centered Capability Ratio that is less than 1.00, or do we believe the six months of experience that tells us that this process is predictable and capable?

Beyond Capability Confusion

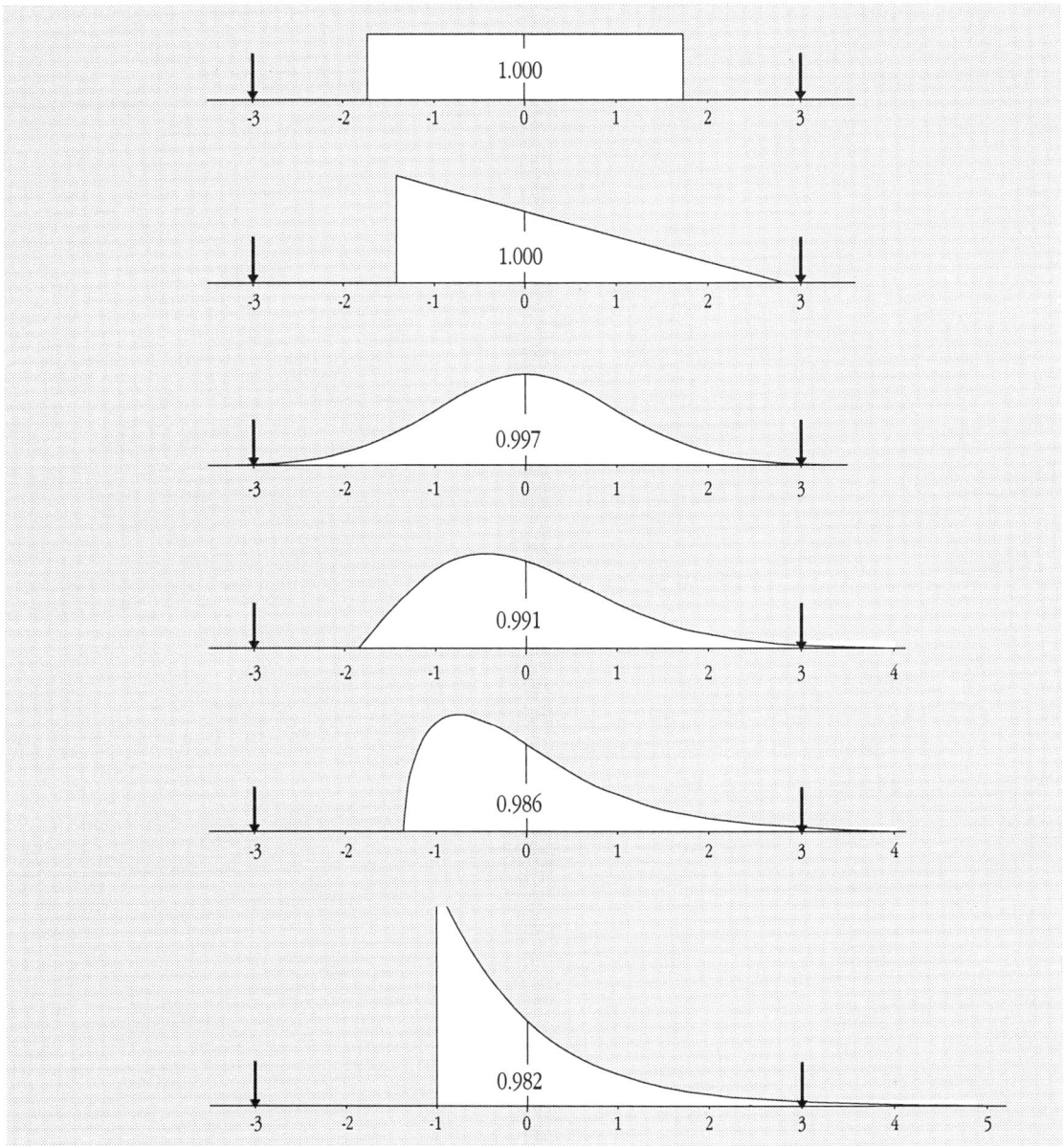

Figure 4.12: The Coverage of Three-Sigma Limits

When a predictable process has capability indexes that are greater than 1.00 the interpretation is clear—the process is capable. But when a predictable process has capability indexes that are less than 1.00 you will need to interpret these values in the light of the past experience with this predictable process. Fortunately, when you have a predictable process, you will have some process history that can be used in this way. In particular, the customer's experience with the product produced may be one source of information. The histogram of the individual values from the process behavior charts would be another source of information about the process. When these sources reveal that your predictable process has been capable, then you should ignore the capability ratio that is less than 1.00.

A capable process can have a capability ratio that is less than 1.00 because of the way in which three-sigma limits characterize the process behavior. Three-sigma limits are generic limits that are intended to work with all types of predictable process behaviors. To see this consider Figure 4.12 which contains six different probability models and their corresponding three-sigma limits.

Three-sigma limits are generic limits that will cover virtually all of the routine variation for a predictable process, no matter what the shape of the histogram of the process outcomes. Which is why you are justified in looking for an explanation for any point that falls outside the three-sigma limits—such points are much more likely to be due to exceptional variation than they are to be due to routine variation.

Consider what happens when a predictable process has a skewed histogram. As can be seen in Figure 4.12, one of the three-sigma limits will cover the bulk of the long tail, while the other limit will tend to be far below the minimum process outcome (or far above the maximum process outcome). While this is an unavoidable consequence of using generic limits, it is not a problem in practice. Skewness in the histogram of the outcomes from a predictable process will tend to occur when the values pile up against a boundary or barrier. (This barrier would be on the left for the four skewed shapes shown in Figure 4.12.) The presence of such a boundary or barrier will create its own *de facto* process limit, and the computed three-sigma limit on the side with the short tail loses any real meaning.

Thus, when we use the computed three-sigma Natural Process Limits in our calculations we are obtaining a generic description of the space needed for a predictable process. Some predictable processes will operate within a smaller interval. This means that the capability indexes are, in effect, worst-case values—the process may well be doing better than the ratio indicates. In Example 4.2, the process was doing much better than the Centered Capability Ratio of 0.86 indicated.

In the case of Example 4.2, where the reason for the skewness is known, and the process has been predictable and capable for the past six months, we could actually modify the way we compute the capability indexes to take this specific process knowledge into account. However, such process knowledge is rare, making the result not worth the effort. The histogram and the process behavior chart already tell the complete story, so why get bogged down in doing calculations. All of the computations we might make will not add one thing to the story which is so eloquently told by these two simple and powerful graphs. Calculations merely complement the graphs—they can never replace them.

For this reason, I prefer to compute the capability indexes in the usual, generic manner, and to use them in concert with the two basic graphs. I prefer to work with standard computations, with their known limitations, than to have to deal with customized and modified calculations which may, or may not, provide a number to characterize a picture, especially when I can easily evaluate that picture my own eyes. While this advice will not sell much software, it will help you to understand and communicate your process capability with greater clarity and integrity than will all of the computations you can dream up and perform.

Be wary of those who go overboard with computations. It is easy to place computations ahead of common sense.

4.4 Another Way to Think About the Capability Indexes

There are many different computational procedures for finding the capability indexes introduced earlier. Since they all result in the same values, none of these is any better or any worse than the others. The following is offered as an alternative in case some readers find it more natural, or more insightful, than the previous description.

The Capability Ratio is based upon the Specified Tolerance, or the Elbow Room for the process. In its raw form this value is expressed in measurement units. However, this value can be converted from measurement units into standard deviation units. This conversion is made by dividing by an adjusted measure of dispersion, *Sigma(X)*, where *Sigma(X)* is defined as:

$$Sigma(X) \;=\; \frac{\overline{s_n}}{c_2} \quad \text{or} \quad \frac{\overline{s}}{c_4} \quad \text{or} \quad \frac{\overline{R}}{d_2} \quad \text{or} \quad \frac{\widetilde{R}}{d_4}$$

It is important to note that these formulas for *Sigma(X)* involve numerators that are within-subgroup measures of dispersion divided by their corresponding bias correction factors, c_2, c_4, d_2, or d_4.

Dividing the Specified Tolerance by *Sigma(X)* will convert the Specified Tolerance from measurement units into standard deviation units. Since the Figure 4.12 tells us that a generic amount of elbow room for any process is six standard deviations, this expression of the Specified Tolerance in standard deviation units becomes immediately interpretable.

Dividing the Distance to Nearer Spec by *Sigma(X)* will also convert this number from measurement units into standard deviation units. When this number is expressed in this way it can be interpreted in the light of Figure 4.12, which tells us that we would like this number to be 3.0 or greater.

Example 4.3: The Reostat Knob Data

For Example 4.1, the Specified Tolerance was 0.030 inches. The Average Range was 0.00856 inches. With a subgroup size of 5 the value for d_2 is 2.326 standard deviations. Thus we get:

$$Sigma(X) \;=\; \frac{\overline{R}}{d_2} \;=\; \frac{0.00856 \text{ inches}}{2.326 \text{ standard deviations}} \;=\; 0.00368 \text{ inches/standard deviation}$$

and so the Specified Tolerance becomes

$$\text{Specified Tolerance} \;=\; 0.030 \text{ inches} \;=\; \frac{0.030 \text{ in.}}{0.00368 \text{ in. / s.d.}} \;=\; 8.15 \text{ standard deviations}$$

Since Figure 4.12 tells us that no process needs more than six standard deviations of elbow room, the fact that the Specified Tolerance in Figure 4.13 is over eight standard deviations wide tells us that we have sufficient elbow room. Finally, this value of 8.15 standard deviations can be converted into the Capability Ratio by dividing by the generic space required (which is six standard deviations):

$$\text{Capability Ratio} \;=\; \frac{8.15 \text{ s.d.}}{6.0 \text{ s.d.}} \;=\; 1.36$$

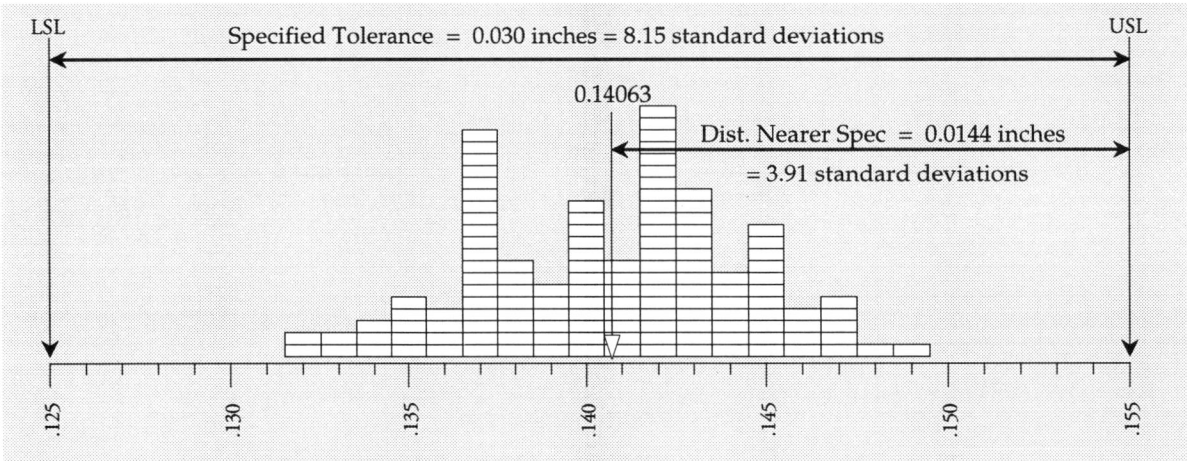

Figure 4.13: The Specified Tolerance and Distance to Nearer Spec for the Rheostat Knobs

For the data of Example 4.1, the Distance to Nearer Spec was 0.0144 in.
Converting this from measurement units into standard deviation units we get:

$$\text{Distance to Nearer Spec} = 0.0144 \text{ in.} = \frac{0.0144 \text{ in.}}{0.00368 \text{ in./s.d.}} = 3.91 \text{ standard deviations}$$

And as before, the Centered Capability Ratio can be obtained by dividing the value above by three standard deviations (which is the required one-sided generic distance shown in Figure 4.12):

$$\text{Centered Capability Ratio} = \frac{3.91 \text{ s.d.}}{3.0 \text{ s.d.}} = 1.30$$

Example 4.4:   **The Ball-Joint Socket Thickness Data**

For the data of Example 4.2 the Specified Tolerance is 15 units, while the Distance to Nearer Spec is 4.66 units. The value for *Sigma(X)* is:

$$Sigma(X) = \frac{3.71 \text{ units}}{2.059 \text{ s.d.}} = 1.80 \text{ units/standard deviation}$$

thus we get a Specified Tolerance of:

$$\text{Specified Tolerance} = 15 \text{ units} = \frac{15 \text{ units}}{1.80 \text{ units/s.d.}} = 8.33 \text{ standard deviations}$$

which gives a Capability Ratio of 8.33/6.0 = 1.39.

In the same manner, the Distance to Nearer Spec is:

$$\text{Distance to Nearer Spec} = 4.66 \text{ units} = \frac{4.66 \text{ units}}{1.80 \text{ units/s.d.}} = 2.59 \text{ standard deviations}$$

which gives us a Centered Capability Ratio of 2.59/3.0 = 0.86.

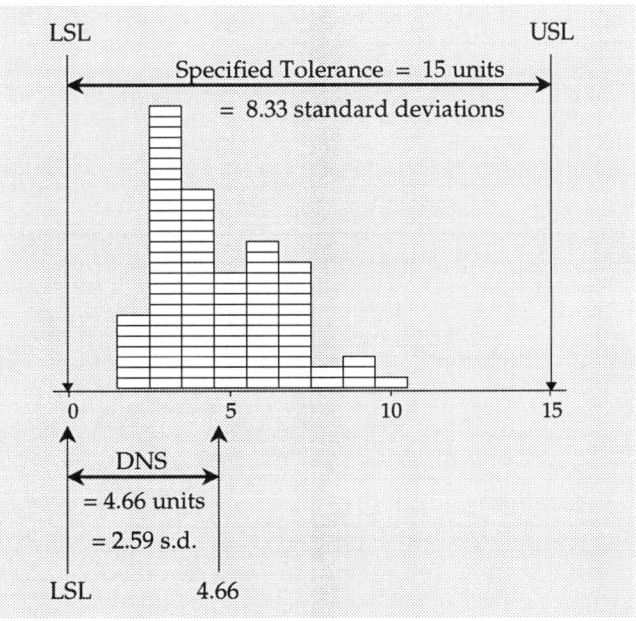

Figure 4.14: The Specified Tolerance and Distance to Nearer Spec for the Ball-Joint Thicknesses

Example 4.5: The Percent Solids Data

A batch of a bulk product is produced every four hours. As each batch is drawn off a sample is obtained and used to characterize that batch. The data below are the percent solids present in the fluid for each of 84 consecutive batches (read these values row by row). Since these values were obtained one-at-a-time they were plotted on a chart for individual values and a moving range (an *XmR* Chart). This chart shows a reasonable degree of predictability for this process.

48	50	49	48	50	50	45	51	47	49	53	51	50	50
47	46	48	47	45	46	47	50	47	49	50	45	50	47
50	49	46	48	46	46	51	48	49	51	48	43	46	48
47	49	49	51	46	47	47	46	50	50	48	49	49	46
48	47	50	48	48	49	47	47	49	48	51	45	48	47
47	49	46	44	48	46	49	50	47	48	47	48	47	47

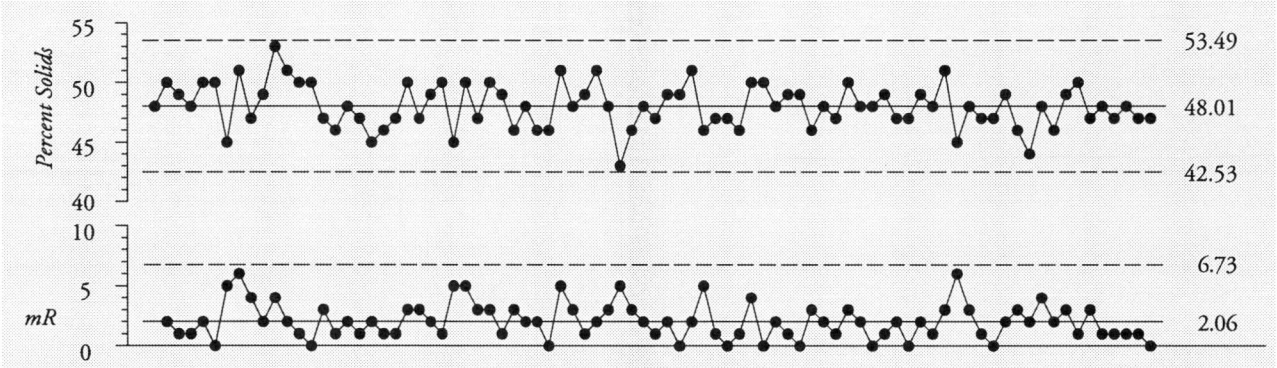

Figure 4.15: The *XmR* Chart for the Percent Solids Data

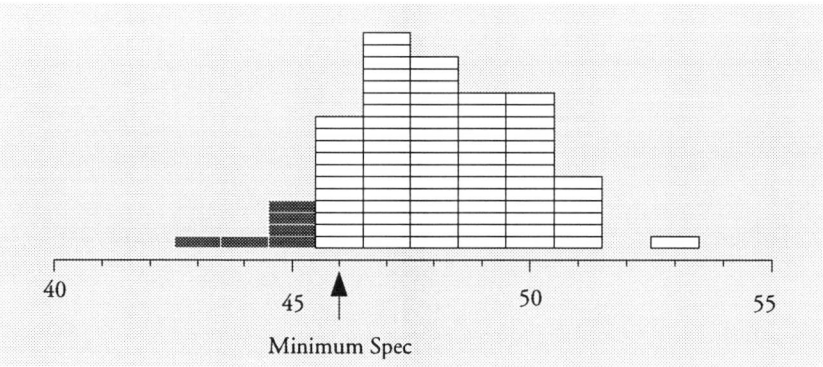

Figure 4.16: The Histogram for the Percent Solids Data

The minimum specification for the percent solids in this product is 46 percent. The histogram of the values from this two week's worth of production shows six values that were out-of-spec. Based on these data, the best estimate of the fraction of nonconforming batches is:

$$\frac{6}{84} = 0.071 \text{ or about 7 percent nonconforming}$$

The Natural Process Limits are the limits shown on the *X* chart. Since this process is currently behaving predictably, and since it has been predictable in the past, we should expect it to continue to produce batches with percent solids values in the range of 43 percent to 53 percent. Thus, the process is predictable, but it is not capable. This process is in the Threshold State.

With a one-sided specification we can only compute the Centered Capability Ratio. The average for these 84 values is 48.01. The minimum specification is effectively 45.5 (46 is acceptable, but 45 is unacceptable). Thus the Distance to Nearer Spec is 48.01 − 45.5 = 2.49 units. Half the distance between the Natural Process Limits is 48.01 − 42.53 = 5.48 units. So the Centered Capability Ratio is:

$$\text{Centered Capability Ratio} = \frac{2.49 \text{ units}}{5.48 \text{ units}} = 0.45$$

which we would interpret to mean that the effective space available for this process is only about 45 percent of the space the process is likely to require.

Using the alternate approach, the *Sigma(X)* value would be:

$$Sigma(X) = \frac{2.06 \text{ units}}{1.128 \text{ s.d.}} = 1.83 \text{ units/standard deviation}$$

Dividing the *DNS* value by *Sigma(X)* will give:

$$\text{Distance to Nearer Spec} = \frac{2.49 \text{ units}}{1.83 \text{ units/s.d.}} = 1.36 \text{ standard deviations}$$

which is much less than the generic requirement of three standard deviations. Thus, the Centered Capability Ratio would be:

$$\text{Centered Capability Ratio} = \frac{1.36 \text{ std. dev.}}{3.0 \text{ std. dev.}} = 0.45$$

This process is in the Threshold State. The histogram and the Centered Capability Ratio convey the same message—some nonconforming batches will be produced. So what can the producer do? Until the process is changed, or the specification is changed, nonconforming batches will continue to occur, and the producer will have to continue to test every batch. If he chooses to work on changing the specifications, he can use the knowledge that his predictable process can reliably yield at least 43 percent solids. If he chooses to change the process he has two options: increase the average percent solids to 51 percent, or reduce the variation from batch to batch.

4.5 Converting Capability Ratios into Fractions Nonconforming

When people began to use capability indexes they had a problem of how to use these strange new numbers to communicate. Just what does it mean to have a Capability Ratio of 1.39 and a Centered Capability Ratio of 0.86 as found in Example 4.2? As was noted there, the Capability Ratio of 1.39 would be interpreted to mean that the space available is about 39 percent larger than the space required for this process. The Centered Capability Ratio of 0.86 would be interpreted to mean that because the process is off-center the effective space available is about 86 percent of the generic space required.

While this interpretation of the capability indexes is simple and straightforward, there is a tendency to want to convert them into something that is more familiar to most people, and this is the fraction nonconforming. In order to make any conversion of capability indexes into fractions of nonconforming product you will have to use some probability distribution in the conversion. This means that you will have to make some assumption about the distribution of the data. The traditional assumption is that the data are normally distributed. Under this assumption we can compute the values shown in Table 4.1.

Table 4.1: Converting Capability Indexes into Percent Nonconforming Using a Normal Distribution

C_p values

C_{pk}	1.10	1.05	1.00	0.95	0.90	0.85	0.80	0.78	0.76	0.74	0.72	0.70	0.68	0.66	0.64	0.62	0.60	0.58
1.10	0.1																	
1.05	0.1	0.2																
1.00	0.2	0.2	0.3															
0.95	0.2	0.2	0.3	0.4														
0.90	0.4	0.4	0.4	0.5	0.7													
0.85	0.5	0.5	0.6	0.6	0.8	1.1												
0.80	0.8	0.8	0.8	0.9	1.0	1.2	1.6											
0.78	1.0	1.0	1.0	1.0	1.1	1.3	1.7	1.9										
0.76	1.1	1.1	1.1	1.2	1.2	1.4	1.7	2.0	2.3									
0.74	1.3	1.3	1.3	1.3	1.4	1.5	1.8	2.0	2.3	2.6								
0.72	1.5	1.5	1.5	1.6	1.6	1.7	2.0	2.1	2.4	2.7	3.1							
0.70	1.8	1.8	1.8	1.8	1.8	1.9	2.1	2.3	2.5	2.8	3.1	3.6						
0.68	2.1	2.1	2.1	2.1	2.1	2.2	2.4	2.5	2.7	2.9	3.2	3.6	4.1					
0.66	2.4	2.4	2.4	2.4	2.4	2.5	2.6	2.7	2.9	3.1	3.3	3.7	4.2	4.8				
0.64	2.7	2.7	2.7	2.8	2.8	2.8	2.9	3.0	3.2	3.3	3.6	3.9	4.3	4.8	5.5			
0.62	3.1	3.1	3.1	3.2	3.2	3.2	3.3	3.4	3.5	3.6	3.8	4.1	4.5	4.9	5.5	6.3		
0.60	3.6	3.6	3.6	3.6	3.6	3.6	3.7	3.8	3.9	4.0	4.2	4.4	4.7	5.1	5.7	6.3	7.2	
0.58	4.1	4.1	4.1	4.1	4.1	4.1	4.2	4.3	4.3	4.4	4.6	4.8	5.1	5.4	5.9	6.5	7.2	8.2
0.56	4.6	4.6	4.6	4.7	4.7	4.7	4.7	4.8	4.8	4.9	5.1	5.2	5.5	5.8	6.2	6.7	7.4	8.2
0.54	5.3	5.3	5.3	5.3	5.3	5.3	5.3	5.4	5.4	5.5	5.6	5.8	6.0	6.2	6.6	7.0	7.6	8.4
0.52	5.9	5.9	5.9	5.9	5.9	6.0	6.0	6.0	6.1	6.1	6.2	6.4	6.5	6.8	7.1	7.5	8.0	8.7
0.50	6.7	6.7	6.7	6.7	6.7	6.7	6.7	6.8	6.8	6.8	6.9	7.0	7.2	7.4	7.6	8.0	8.5	9.1
0.48	7.5	7.5	7.5	7.5	7.5	7.5	7.5	7.6	7.6	7.6	7.7	7.8	7.9	8.1	8.3	8.6	9.0	9.6
0.46	8.4	8.4	8.4	8.4	8.4	8.4	8.4	8.4	8.5	8.5	8.5	8.6	8.7	8.9	9.1	9.3	9.7	10.2
0.44	9.3	9.3	9.3	9.3	9.3	9.3	9.4	9.4	9.4	9.4	9.5	9.5	9.6	9.8	9.9	10.2	10.5	10.9
0.42	10.4	10.4	10.4	10.4	10.4	10.4	10.4	10.4	10.4	10.5	10.5	10.5	10.6	10.7	10.9	11.1	11.3	11.7
0.40	11.5	11.5	11.5	11.5	11.5	11.5	11.5	11.5	11.5	11.6	11.6	11.6	11.7	11.8	11.9	12.1	12.3	12.6
0.38	12.7	12.7	12.7	12.7	12.7	12.7	12.7	12.7	12.7	12.8	12.8	12.8	12.9	13.0	13.1	13.2	13.4	13.7
0.36	14.0	14.0	14.0	14.0	14.0	14.0	14.0	14.0	14.0	14.0	14.1	14.1	14.1	14.2	14.3	14.4	14.6	14.8
0.34	15.4	15.4	15.4	15.4	15.4	15.4	15.4	15.4	15.4	15.4	15.4	15.5	15.5	15.6	15.6	15.7	15.9	16.1
0.32	16.9	16.9	16.9	16.9	16.9	16.9	16.9	16.9	16.9	16.9	16.9	16.9	16.9	17.0	17.1	17.1	17.3	17.4
0.30	18.4	18.4	18.4	18.4	18.4	18.4	18.4	18.4	18.4	18.4	18.4	18.5	18.5	18.5	18.6	18.6	18.8	18.9
0.28	20.0	20.0	20.0	20.0	20.0	20.0	20.0	20.1	20.1	20.1	20.1	20.1	20.1	20.1	20.2	20.2	20.3	20.5
0.26	21.8	21.8	21.8	21.8	21.8	21.8	21.8	21.8	21.8	21.8	21.8	21.8	21.8	21.8	21.9	21.9	22.0	22.1
0.24	23.6	23.6	23.6	23.6	23.6	23.6	23.6	23.6	23.6	23.6	23.6	23.6	23.6	23.6	23.7	23.7	23.8	23.9
0.22	25.5	25.5	25.5	25.5	25.5	25.5	25.5	25.5	25.5	25.5	25.5	25.5	25.5	25.5	25.5	25.6	25.6	25.7
0.20	27.4	27.4	27.4	27.4	27.4	27.4	27.4	27.4	27.4	27.4	27.4	27.4	27.5	27.5	27.5	27.5	27.6	27.6
0.18	29.5	29.5	29.5	29.5	29.5	29.5	29.5	29.5	29.5	29.5	29.5	29.5	29.5	29.5	29.5	29.5	29.6	29.6
0.16	31.6	31.6	31.6	31.6	31.6	31.6	31.6	31.6	31.6	31.6	31.6	31.6	31.6	31.6	31.6	31.6	31.7	31.7
0.14	33.7	33.7	33.7	33.7	33.7	33.7	33.7	33.7	33.7	33.7	33.7	33.7	33.7	33.7	33.8	33.8	33.8	33.8
0.12	35.9	35.9	35.9	35.9	35.9	35.9	35.9	35.9	35.9	35.9	35.9	35.9	36.0	36.0	36.0	36.0	36.0	36.0
0.10	38.2	38.2	38.2	38.2	38.2	38.2	38.2	38.2	38.2	38.2	38.2	38.2	38.2	38.2	38.2	38.2	38.3	38.3
0.08	40.5	40.5	40.5	40.5	40.5	40.5	40.5	40.5	40.5	40.5	40.5	40.5	40.5	40.5	40.5	40.5	40.6	40.6
0.06	42.9	42.9	42.9	42.9	42.9	42.9	42.9	42.9	42.9	42.9	42.9	42.9	42.9	42.9	42.9	42.9	42.9	42.9
0.04	45.2	45.2	45.2	45.2	45.2	45.2	45.2	45.2	45.2	45.2	45.2	45.2	45.2	45.2	45.2	45.2	45.2	45.3
0.02	47.6	47.6	47.6	47.6	47.6	47.6	47.6	47.6	47.6	47.6	47.6	47.6	47.6	47.6	47.6	47.6	47.6	47.6
0.00	50.0	50.0	50.0	50.0	50.0	50.0	50.0	50.0	50.0	50.0	50.0	50.0	50.0	50.0	50.0	50.0	50.0	50.0

Table 4.1: Converting Capability Indexes into Percent Nonconforming Using a Normal Distribution

C_p values

C_{pk}	0.56	0.54	0.52	0.50	0.48	0.46	0.44	0.42	0.40	0.38	0.36	0.34	0.32	0.30	0.28	0.26	0.24
0.56	9.3																
0.54	9.4	10.5															
0.52	9.5	10.6	11.9														
0.50	9.8	10.8	11.9	13.4													
0.48	10.2	11.1	12.1	13.4	15.0												
0.46	10.8	11.5	12.5	13.6	15.1	16.8											
0.44	11.4	12.1	12.9	14.0	15.3	16.8	18.7										
0.42	12.2	12.8	13.5	14.5	15.6	17.1	18.8	20.8									
0.40	13.0	13.6	14.2	15.1	16.2	17.4	19.0	20.8	23.0								
0.38	14.0	14.5	15.1	15.9	16.8	18.0	19.4	21.1	23.1	25.4							
0.36	15.1	15.5	16.1	16.8	17.6	18.7	19.9	21.5	23.3	25.5	28.0						
0.34	16.4	16.7	17.2	17.8	18.5	19.5	20.6	22.1	23.8	25.8	28.1	30.8					
0.32	17.7	18.0	18.4	18.9	19.6	20.4	21.5	22.8	24.3	26.2	28.4	30.9	33.7				
0.30	19.1	19.4	19.7	20.2	20.8	21.6	22.5	23.7	25.1	26.8	28.8	31.1	33.8	36.8			
0.28	20.6	20.9	21.2	21.6	22.1	22.8	23.6	24.7	26.0	27.5	29.4	31.6	34.1	36.9	40.1		
0.26	22.3	22.5	22.7	23.1	23.6	24.2	24.9	25.9	27.0	28.5	30.1	32.2	34.5	37.2	40.2	43.5	
0.24	24.0	24.2	24.4	24.7	25.1	25.6	26.3	27.2	28.2	29.5	31.1	32.9	35.1	37.6	40.4	43.6	47.2
0.22	25.8	26.0	26.2	26.4	26.8	27.2	27.8	28.6	29.6	30.7	32.1	33.8	35.8	38.2	40.8	43.9	47.2
0.20	27.7	27.8	28.0	28.2	28.6	29.0	29.5	30.2	31.0	32.1	33.4	34.9	36.8	38.9	41.4	44.3	47.5
0.18	29.7	29.8	30.0	30.2	30.4	30.8	31.2	31.8	32.6	33.6	34.7	36.1	37.8	39.8	42.2	44.8	47.9
0.16	31.8	31.9	32.0	32.1	32.4	32.7	33.1	33.6	34.3	35.2	36.2	37.5	39.1	40.9	43.1	45.6	48.4
0.14	33.9	34.0	34.1	342	34.4	34.7	35.0	35.5	36.1	36.9	37.8	39.0	40.4	42.1	44.1	46.4	49.1
0.12	36.1	36.1	36.2	36.4	36.5	36.8	37.1	37.5	38.0	38.7	39.5	40.6	41.9	43.4	45.3	47.4	49.9
0.10	38.3	38.4	38.4	38.6	38.7	38.9	39.2	39.5	40.0	40.6	41.4	42.3	43.5	44.9	46.6	48.6	50.9
0.08	40.6	40.7	40.7	40.8	40.9	41.1	41.3	41.6	42.1	42.6	43.3	44.1	45.2	46.5	48.0	49.9	52.0
0.06	42.9	43.0	43.0	43.1	43.2	43.4	43.6	43.8	44.2	44.6	45.2	46.0	47.0	48.1	49.5	51.2	53.2
0.04	45.3	45.3	45.4	45.4	45.5	45.6	45.8	46.0	46.4	46.8	47.3	48.0	48.8	49.9	51.2	52.7	54.6
0.02	47.7	47.7	47.7	47.8	47.8	48.0	48.1	48.3	48.6	48.9	49.4	50.0	50.8	51.7	52.9	54.3	56.0
0.00	50.0	50.1	50.1	50.1	50.2	50.3	50.4	50.6	50.8	51.1	51.5	52.1	52.7	53.6	54.6	55.9	57.5

C_p values

C_{pk}	0.22	0.20	0.18	0.16	0.14	0.12	0.10	0.08	0.06	0.04	0.02	0.00
0.22	50.9											
0.20	51.0	54.9										
0.18	51.2	54.9	58.9									
0.16	51.6	55.1	59.0	63.1								
0.14	52.1	55.5	59.2	63.2	67.4							
0.12	52.8	56.0	59.5	63.4	67.5	71.9						
0.10	53.6	56.6	60.0	63.7	67.7	71.9	76.4					
0.08	54.5	57.4	60.6	64.1	67.9	72.1	76.5	81.0				
0.06	55.6	58.2	61.3	64.6	68.3	72.3	76.6	81.1	85.7			
0.04	56.7	59.2	62.1	65.3	68.8	72.6	76.8	81.2	85.7	90.4		
0.02	58.0	60.3	63.0	66.0	69.4	73.1	77.1	81.3	85.8	90.5	95.2	
0.00	59.3	61.5	64.0	66.9	70.0	73.6	77.4	81.6	85.9	90.5	95.2	100.0

Table 4.1: Converting Capability Indexes into Percent Nonconforming Using a Normal Distribution

	C_p values																
C_{pk}	1.10	0.48	0.44	0.40	0.36	0.32	0.28	0.24	0.20	0.18	0.16	0.14	0.12	0.10	0.08	0.06	0.04	0.02
0.00	50.0	50.2	50.4	50.8	51.5	52.7	54.6	57.5	61.5	64.0	66.9	70.0	73.6	77.4	81.6	85.9	90.5	95.2
−0.05	56.0	56.1	56.2	56.5	57.0	57.9	59.3	61.6	64.8	66.9	69.3	72.1	75.2	78.6	82.4	86.5	90.8	95.3
−0.10	61.8	61.9	62.0	62.1	62.5	63.1	64.2	65.9	68.5	70.2	72.2	74.5	77.2	80.2	83.6	87.3	91.3	95.5
−0.15	67.4	67.4	67.5	67.6	67.8	68.3	69.0	70.3	72.3	73.7	75.3	77.2	79.5	82.1	85.0	88.3	91.9	95.8
−0.20	72.6	72.6	72.6	72.7	72.9	73.2	73.7	74.6	76.2	77.2	78.5	80.1	81.9	84.1	86.6	89.4	92.6	96.2
−0.25	77.3	77.4	77.4	77.4	77.5	77.7	78.1	78.8	79.9	80.7	81.7	82.9	84.4	86.2	88.3	90.7	93.4	96.6
−0.30	81.6	81.6	81.6	81.6	81.7	81.8	82.1	82.6	83.4	84.0	84.7	85.7	86.9	88.3	90.0	92.0	94.3	97.0
−0.35	85.3	85.3	85.3	85.3	85.4	85.5	85.6	86.0	86.5	87.0	87.5	88.3	89.2	90.3	91.6	93.2	95.2	97.4
−0.40	88.5	88.5	88.5	88.5	88.5	88.6	88.7	88.9	89.3	89.6	90.0	90.6	91.2	92.1	93.1	94.4	96.0	97.8
−0.45	91.2	91.2	91.2	91.2	91.2	91.2	91.3	91.4	91.7	91.9	92.2	92.6	93.1	93.7	94.5	95.5	96.7	98.2
−0.50	93.3	93.3	93.3	93.3	93.3	93.4	93.4	93.5	93.7	93.8	94.0	94.3	94.6	95.1	95.7	96.5	97.4	98.6
−0.55	95.1	95.1	95.1	95.1	95.1	95.1	95.1	95.2	95.3	95.4	95.5	95.7	95.9	96.3	96.7	97.3	98.0	98.9
−0.60	96.4	96.4	96.4	96.4	96.4	96.4	96.4	96.5	96.5	96.6	96.7	96.8	97.0	97.2	97.5	97.9	98.5	99.1
−0.65	97.4	97.4	97.4	97.4	97.4	97.4	97.5	97.5	97.5	97.6	97.6	97.7	97.8	98.0	98.2	98.5	98.9	99.4
−0.70	98.2	98.2	98.2	98.2	98.2	98.2	98.2	98.2	98.3	98.3	98.3	98.4	98.5	98.6	98.7	98.9	99.2	99.5
−0.75	98.8	98.8	98.8	98.8	98.8	98.8	98.8	98.8	98.8	98.8	98.8	98.9	98.9	99.0	99.1	99.2	99.4	99.7
−0.80	99.2	99.2	99.2	99.2	99.2	99.2	99.2	99.2	99.2	99.2	99.2	99.2	99.3	99.3	99.4	99.5	99.6	99.8
−0.85	99.5	99.5	99.5	99.5	99.5	99.5	99.5	99.5	99.5	99.5	99.5	99.5	99.5	99.5	99.6	99.6	99.7	99.8
−0.90	99.7	99.7	99.7	99.7	99.7	99.7	99.7	99.7	99.7	99.7	99.7	99.7	99.7	99.7	99.7	99.8	99.8	99.8
−0.95	99.8	99.8	99.8	99.8	99.8	99.8	99.8	99.8	99.8	99.8	99.8	99.8	99.8	99.8	99.8	99.8	99.9	99.9
−1.00	99.9	99.9	99.9	99.9	99.9	99.9	99.9	99.9	99.9	99.9	99.9	99.9	99.9	99.9	99.9	99.9	99.9	100.0
−1.05	99.9	99.9	99.9	99.9	99.9	99.9	99.9	99.9	99.9	99.9	99.9	99.9	99.9	99.9	99.9	99.9	100.0	
−1.10	100.0	100.0	100.0	100.0	100.0	100.0	100.0	100.0	100.0	100.0	100.0	100.0	100.0	100.0	100.0	100.0		

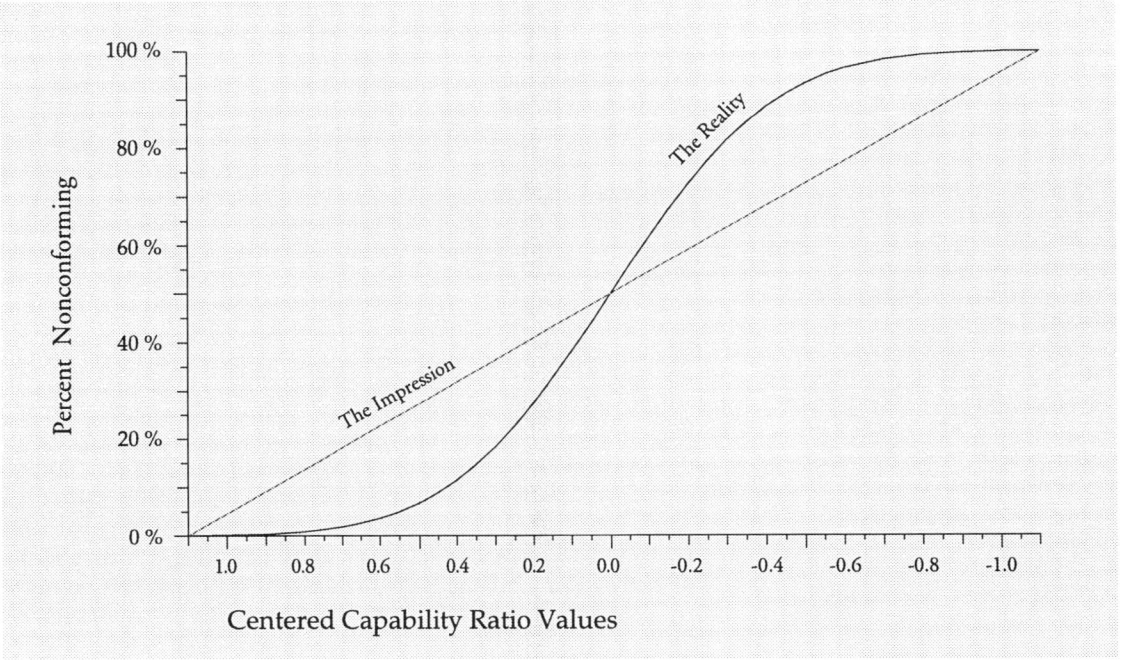

Figure 4.17: The Nonlinear Relationship Between C_{pk} and Fraction Nonconforming

Even a brief examination of the values in Table 4.1 will show the extreme nonlinear nature of the capability indexes. At different places in the table a one-unit change in either of the capability indexes will result in a different amount of change in the percent nonconforming. At first the percentage of nonconforming product increases very slowly as the capability index goes down. Later it increases more rapidly. This lack of linearity means that things will generally sound worse than they really are when you use the capability indexes to characterize a process.

Figure 4.17 shows this nonlinear relationship between the Centered Capability Ratio and the percent nonconforming for the case where C_p = 1.10. This nonlinear relationship is important because our natural inclination is to interpret things in a linear fashion. That is, we expect a change in the capability index from 1.0 to 0.9 to mean the same thing as a change from 0.6 to 0.5. Since this is not the case it is easy to misinterpret the capability indexes. Converting them into estimated percent nonconforming values will help to prevent this mistake.

You should also note that Table 4.1 covers virtually every situation between 1 part per thousand nonconforming and 1000 parts per thousand nonconforming. By the time you have capability indexes larger than 1.10 you are in the realm of 100 parts per million or less. (And if your Centered Capability Ratio should ever drop below –1.10 you will have fewer than 100 parts per million that are conforming.) This is why we like to have capability indexes that are slightly greater than 1.00. So if you have capability indexes that are not covered by Table 4.1, then your predictable process is either very very good or very very bad. (You should not need to perform any calculations to figure out which!)

Table 4.1 is based on the normal probability distribution because so many different phenomena are reasonably modeled by this one distribution. It provides ball-park estimates of the fraction nonconforming, and that is about all that you really need. While the computations behind Table 4.1 could be performed using other probability distributions, and while some software packages attempt to perform this custom type of conversion, there is little to be gained by such efforts. If your capability indexes place you somewhere within Table 4.1, then you can estimate the fraction nonconforming in the much more direct manner outlined below. If your capability indexes place you outside Table 4.1 then see the next section.

No matter what your capability indexes may be, the best estimate of the fraction of nonconforming product will always be the simple ratio:

$$\text{Fraction Nonconforming} = \frac{\text{total number nonconforming found}}{\text{total number examined}}$$

- No fancy computer program is needed.
- No esoteric probability distribution functions are needed.
- No assumptions about the normality, or nonnormality, of the data are needed.
- Simply count the number nonconforming and divide by the total number examined.

This also applies to the conversion of capability indexes into fraction nonconforming. Any careful computation of capability indexes will have begun with a process behavior chart. The data on the process behavior chart can be placed in a histogram with the specification limits superimposed. This histogram can then be used with the formula above to estimate the fraction nonconforming directly—no assumptions needed. If the process is predictable, then this fraction nonconforming will be a reasonable prediction of what you can expect from this process. If the process is unpredictable, then *no* estimate of the fraction nonconforming will be predictive of the future. With an unpredictable process the ratio above will merely provide a description of the past.

So, rather than trying to convert a capability ratio into a fraction nonconforming, use the same data that was used to generate the capability ratio, and compute the fraction nonconforming directly.

When a capability ratio is used in place of the fraction nonconforming as a way to judge (or to con-

Beyond Capability Confusion

demn) a production process this nonlinear relationship between the C_{pk} statistic and the fraction nonconforming tends to make things sound worse than they really are.

Recently many software packages have developed fancy routines to fit a probability distribution to the histogram of the data in order to "better convert capability indexes into fractions nonconforming." This is an example of computation winning out over common sense. In most cases the uncertainty in the fraction nonconforming will be greater than the refinement offered by such a conversion. Extreme examples of this sort of nonsense occur when capability indexes which are greater than 1.00 are converted into parts per million defect rates.

4.6 The Fraction Nonconforming When Capability Indexes Exceed 1.00

As can be seen in Table 4.1, whenever the capability indexes exceed 1.00 you will have a very small theoretical probability of nonconforming product. While the values in Table 4.1 were based on the normal probability model, the previous statement will be true for virtually any realistic probability model you might choose. We saw in Figure 4.12 that three-sigma limits will cover virtually all of the routine variation of a predictable process, leaving very little to fall outside. And this is true regardless of the shape of the probability distribution used. So if we try to convert capability indexes that are greater than 1.00 into an estimated fraction nonconforming we will have to evaluate the *infinitesimal areas* under the *extreme tails* of a *probability distribution* that we have *assumed* to be the correct one for our predictable process.

Using the traditional normal distribution, and assuming that the Centered Capability Ratio is equal to the Capability Ratio for simplicity, we could compute the following values.

Capability indexes of 1.00 would be converted into 2700 parts per million nonconforming.
Capability indexes of 1.10 would be converted into 968 parts per million nonconforming.
Capability indexes of 1.17 would be converted into 466 parts per million nonconforming.
Capability indexes of 1.33 would be converted into 63 parts per million nonconforming.
Capability indexes of 1.50 would be converted into 6.8 parts per million nonconforming.
Capability indexes of 1.67 would be converted into 0.574 parts per million nonconforming.
Capability indexes of 1.83 would be converted into 0.038 parts per million nonconforming.
Capability indexes of 2.00 would be converted into 0.002 parts per million nonconforming.

While you can certainly compute these vary small probabilities, there is a problem with trying to use them. Even if your assumed distribution is appropriate, and even if you know, or can compute, the infinitesimal areas under the extreme tails of your assumed distribution, your results are still going to be approximate. The probability model does not govern your process. The probability model does not generate your data. The probability model which you have assumed and used is only an approximation to the behavior of your process. Therefore, you should not expect it to *exactly* describe your process outcomes. In particular, one of the common discrepancies between a probability model and the behavior of a predictable process will be in the extreme tails. Probability models often have tails that run all the way out to infinity, while histograms always have finite tails. This means that even the best of approximations will part company with your histogram somewhere out beyond the three-sigma limits. And the infinitesimal areas under the extreme tails of the probability model will no longer characterize your process.

And then there is the problem of choosing a probability model. To illustrate the sensitivity of these tail areas to the choice of probability model we will compare the normal distribution with a Burr distribution that is essentially the same as the normal.

The Burr distribution has a cumulative distribution function of:

$$F(x) = 1 - (1 + x^c)^{-k} \quad \text{for } x > 0$$

Standardizing this distribution gives a probability density function of:

$$f(z) = \sigma c k \, [\, 1 + (\mu + \sigma z)^c \,]^{-k-1} \, (\mu + \sigma z)^{c-1} \quad \text{for } z > -\frac{\mu}{\sigma}$$

With $c = 4.873717$ and $k = 6.157568$, this distribution has a mean of $\mu = 0.644717$, a standard deviation of $\sigma = 0.161990$, a skewness of $\alpha_3 = 0.00$ and a kurtosis of $\alpha_4 = 3.00$. Thus, the standardized form of this distribution has a mean of 0.00, a standard deviation of 1.00, a skewness of 0.00, and a kurtosis of 3.00. In other words, this standardized Burr distribution closely approximates the standard normal distribution, as shown in Figure 4.18.

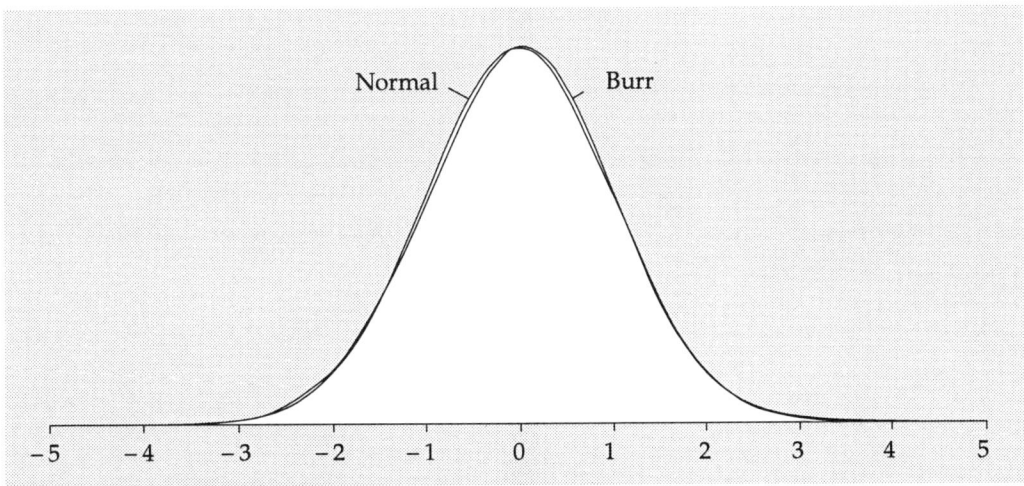

Figure 4.18: Comparison of a Standardized Burr and the Standard Normal Distributions

You will never have enough data to determine which of the two curves in Figure 4.18 might provide the best approximation to your data.

Yet the Burr distribution in Figure 4.18 would result in the following fractions nonconforming. If both the C_p and C_{pk} values are 1.00, then the fraction nonconforming would be 0.002473, or 2473 ppm. If both the C_p and C_{pk} values are 1.33, then the fraction nonconforming would be 96 ppm. If both the C_p and C_{pk} values are 1.50, then the fraction nonconforming would be 22 ppm. And if both the C_p and C_{pk} values are 2.00, then the fraction nonconforming would be 0.310 ppm.

Table 4.2 Fractions Nonconforming Under The Extreme Tails

Capability Indexes	Normal Distribution		Burr Distribution
1.00	2700 ppm	>	2473 ppm
1.33	63 ppm	<	96 ppm
1.50	6.8 ppm	<	22 ppm
2.00	.002 ppm	<	.310 ppm

Since you will never have enough data to "choose" between the two curves above, you will have to *assume* that one of them is correct. Your calculations will be no better than your assumption. Do you really want to calculate values out to *nine decimal places* based on your *assumption*?

Recall Shewhart's Second Rule for the presentation of data:
"Any summary of a distribution...
    should not give an objective degree of belief
        in any inference or prediction...
            that would cause human action
                significantly different from what this action would be
                    if the original data had been taken as a basis for evidence."

The best estimate of the fraction nonconforming is still the ratio shown in the previous section. Any fraction nonconforming computed from a theoretical distribution which corresponds to areas beyond three sigma is purely imaginary. Such numbers have no basis in reality. They serve no practical purpose—except to impress the numerically naive.

4.7 Summary

Thus, capability involves both the predictability of the process and the conformance of product to the specifications. While capability indexes may be used as part of the evaluation of the capability of a predictable process, they cannot be used alone—they need the context provided by the process behavior chart and the histogram of the individual values.

Chapter Five

What Can Be Said for Unpredictable Processes?

Not much.

An unpredictable process is one that has failed to display any reasonable degree of consistency in the past. It is therefore illogical to expect that such a process will spontaneously begin to behave consistently in the future.

Read the previous paragraph again.

What part of unpredictable do you not understand?

5.1 Unpredictable Processes

The very unpredictability of an unpredictable process will undermine any attempt to define the capability of that process. While you may use the specifications to characterize the past *outcomes*, and while you may compute the percentage of nonconforming outcomes in the past, this percentage is not the same as a characterization of the *process*. Even though you may know the percentage of nonconforming outcomes in the past, the lack of a reasonable degree of consistency in the process will prevent you from using the past to characterize the future. And process capability is about the future. (You certainly would not be reading this book if all you wanted to know was how to compute the percent nonconforming in the past!)

"But my customer is pressuring me to provide him with capability indexes for a process that is unpredictable. What can I do? Can I calculate the capability indexes anyway?" The following examples will illustrate the consequences of this course of action.

Example 5.1: The Batch Weight Data

A particular product is produced in batches. Each batch is blended in a pharmaceutical blender and the final weight of each batch is recorded. While a shipment consists of 30 batches, the batch identity is preserved by the subsequent processing. And, of course, the customer wants the supplier to provide the capability indexes for each shipment. The data for one shipment are:

Weights	905	930	865	895	905	885	890	930	915	910	920	915	925	860	905
mov. Ranges	—	25	65	30	10	20	5	40	15	5	10	5	10	65	45
Weights	925	925	905	915	930	890	940	860	875	985	970	940	975	1000	1035
mov. Ranges	20	0	20	10	15	40	50	80	15	110	15	30	35	25	35

The average weight of these 30 batches is 920.8 kg. The median moving range is 20 kg. For purposes of this example assume that the specifications for these weights are 850 kg to 990 kg. The median moving range is converted into a value for *Sigma(X)* by dividing it by the bias correction factor d_4:

$$Sigma(X) = \frac{20 \text{ kg}}{0.954 \text{ std. dev.}} = 20.96 \text{ kg/s.d.}$$

Using this value, the Natural Process Limits are found to be:

$$920.8 \text{ kg} \pm 62.9 \text{ kg} = 857.9 \text{ kg to } 983.7 \text{ kg}$$

Thus, the Elbow Room is 990 − 850 = 140 kg, which is the space available for this process. The distance between the Natural Process Limits is 983.7 − 857.9 = 125.8 kg, which is the space that this process should require. Combining these we get a Capability Ratio of:

$$C_p = \frac{140 \text{ units}}{125.8 \text{ units}} = 1.11$$

The customer will probably interpret this value to mean that the space available is about 11 percent larger than the space required.

The Distance to the Nearer Spec will characterize the process location relative to the specifications. Here the *DNS* is 990 − 920.8 = 69.2 kg. Half of the distance between the Natural Process Limits is 62.9 kg. Thus the Centered Capability Ratio is:

$$C_{pk} = \frac{69.2 \text{ units}}{62.9 \text{ units}} = 1.10$$

These capability indexes are not bad. They suggest a process that is capable of meeting the specifications, and that is well centered within the specifications. The image suggested by these indexes is shown in Figure 5.1.

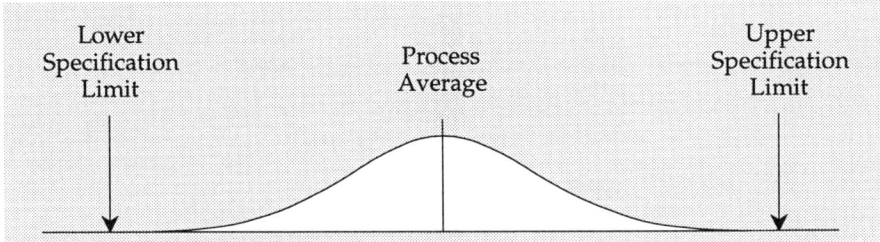

Figure 5.1: How We Visualize a Process Having C_{pk} = 1.10

However, the actual situation is somewhat different than the capability indexes, and Figure 5.1, would lead you to believe. The histogram of the 30 values is shown in Figure 5.2. There you will see that two of the 30 batches have values that are outside the specs. Thus, in the data which gave us capability indexes greater than 1.00 we have about seven percent nonconforming batches. The customer may like the capability indexes, but he will be unhappy with the nonconforming batches.

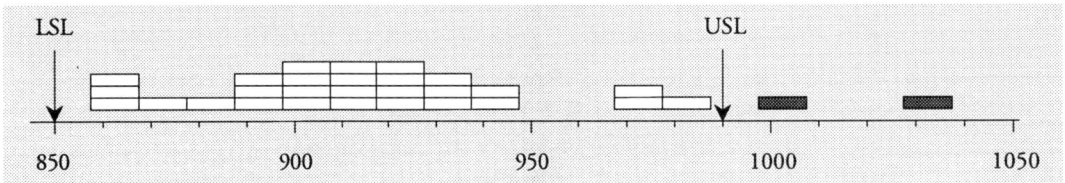

Figure 5.2: Histogram of the 30 Batch Weights Having C_{pk} = 1.10

Of course, the reason for the discrepancy between the capability indexes and the fraction nonconforming comes from the unpredictable nature of this process. The 30 batch weights are shown on an *XmR* chart in Figure 5.3. Three of these batch weights and two of the moving ranges fall outside the limits.

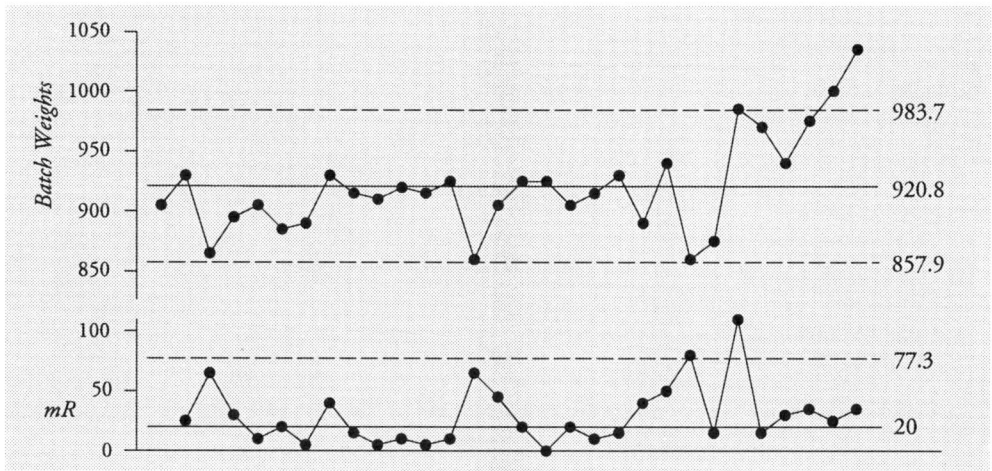

Figure 5.3: XmR Chart for the 30 Batch Weights

So this process is not producing 100 percent conforming product. Can we characterize the amount of nonconforming product that we will expect in the future? We had seven percent nonconforming batch weights in the first shipment. Could you use this historical percentage of nonconforming product as an approximation of what you could expect from this process in the future? Unfortunately, as appealing as this idea is, it will not work. To see why this is so, consider the next 60 batches:

```
1020  985  960  945  965      940  900  920  980  950      955  970  970 1035 1040
1000 1000  990 1000  950      940  965  920  920  925      900  905  900  925  885

1005 1005  950  920  875      865  880  960  925  925      875  900  905  990  970
 910  980  900  970  900      895  885  925  870  875      910  915  900  950  880
```

Eight of the next sixty batch weights are outside the specification limits. This is 13 percent nonconforming, which is twice the fraction nonconforming in the first 30 batches.

"Well, perhaps 30 batches just wasn't enough data. Couldn't we use the fraction nonconforming from the first 90 batches to better characterize this process?" Ten nonconforming batches out of 90 is 11 percent nonconforming. Should we expect about 11 percent nonconforming in the future? The next 90 batch weights were:

```
910  965  910  880  900      920  940  985  965  925      925  975  905  890  950
975  935  940  900  915      980  880  905  915  960      900  915  920  865  980

935  840  900  965  890      875 1020  780  900  900      800  960  845  820  910
885  940  930  925  850      965 1010 1030  980 1010      950  940 1005  880  930

845  935  905  965  975      985  975  950  905  965      905  950  905  995  900
840 1050  935  940  920      985  970  915  935  950     1030  875  880  955  910
```

Fifteen of these 90 batch weights are outside the specifications. This is 17 percent rather than the 11 percent we expected. Oops!

"Well, maybe 90 batches wasn't enough either. Let's use all the data to date. Twenty-five of the 180 batches produced so far have been outside the specs. This is 14 percent. Surely now we have value that is in the right ball-park." The next 79 batch weights were:

```
1050  890 1005  915 1070     970 1040  770  940  950    1040 1035 1110  845  900
 905  910  860 1045  820     900  860  875 1005  880     750  900  835  930  860
 960  950 1020  975  950     960  950  880 1000 1005     990 1020  980 1020  920
 960 1000 1000  860 1130     830  965  930  950  945     900  990  865  945  970
 915  975  940  870  890     915  935 1060 1015 1100     810 1010 1140  805 1020
1110  975  970 1090
```

Of these last 79 batches, 33 were outside the specs, for a total of 42 percent nonconforming. So how did we do in predicting the fraction nonconforming for this process?

Neither the capability ratio of 1.10 computed from the first 30 batches, nor the seven percent nonconforming found in the first 30 batches even come close to describing what this process subsequently produced. Using more data did not remedy this problem. The histogram of all 259 batches produced during this one period is shown in Figure 5.4. In all, a total of 58 of these 259 batch weights were outside the specification limits, for a total of about 22 percent nonconforming. Care to speculate on what the percent nonconforming will be next month?

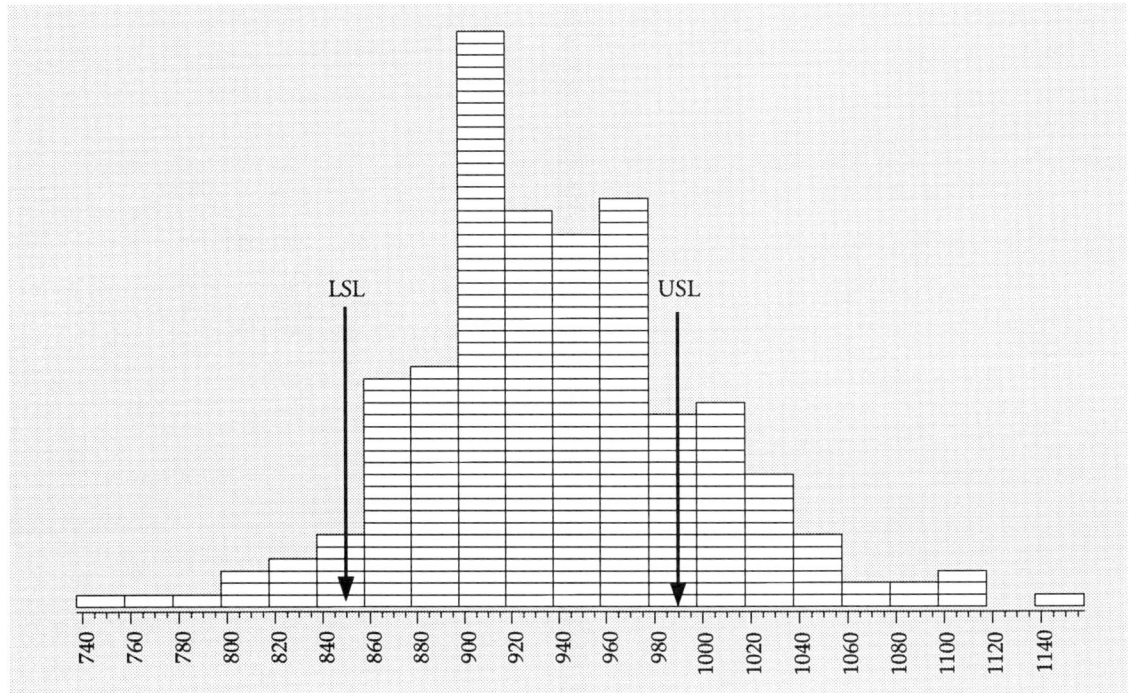

Figure 5.4: Histogram of 259 Batch Weights

The 259 batch weights are shown on an *X* chart in Figure 5.5. The limits on this chart are the limits found from the first 30 values. The upper Natural Process Limit is 983.7, the central line is 920.8, and the lower Natural Process Limit is 857.9. In Figure 5.3 the first 30 values gave us a warning that this process was not predictable, and Figure 5.5 shows that it certainly lived up to this promise.

Now why does all this matter? Can't the customer simply use the compound as he needs it? Why all this concern about the weight of each batch of compound? The first reason that the batch weights are important is that the batch identity must be preserved throughout subsequent operations. The customer cannot mix compound from different batches together. And the second reason has to do with the way the mixing is done. Trace ingredients are trace ingredients because a little bit goes a long way. When mixing this compound the

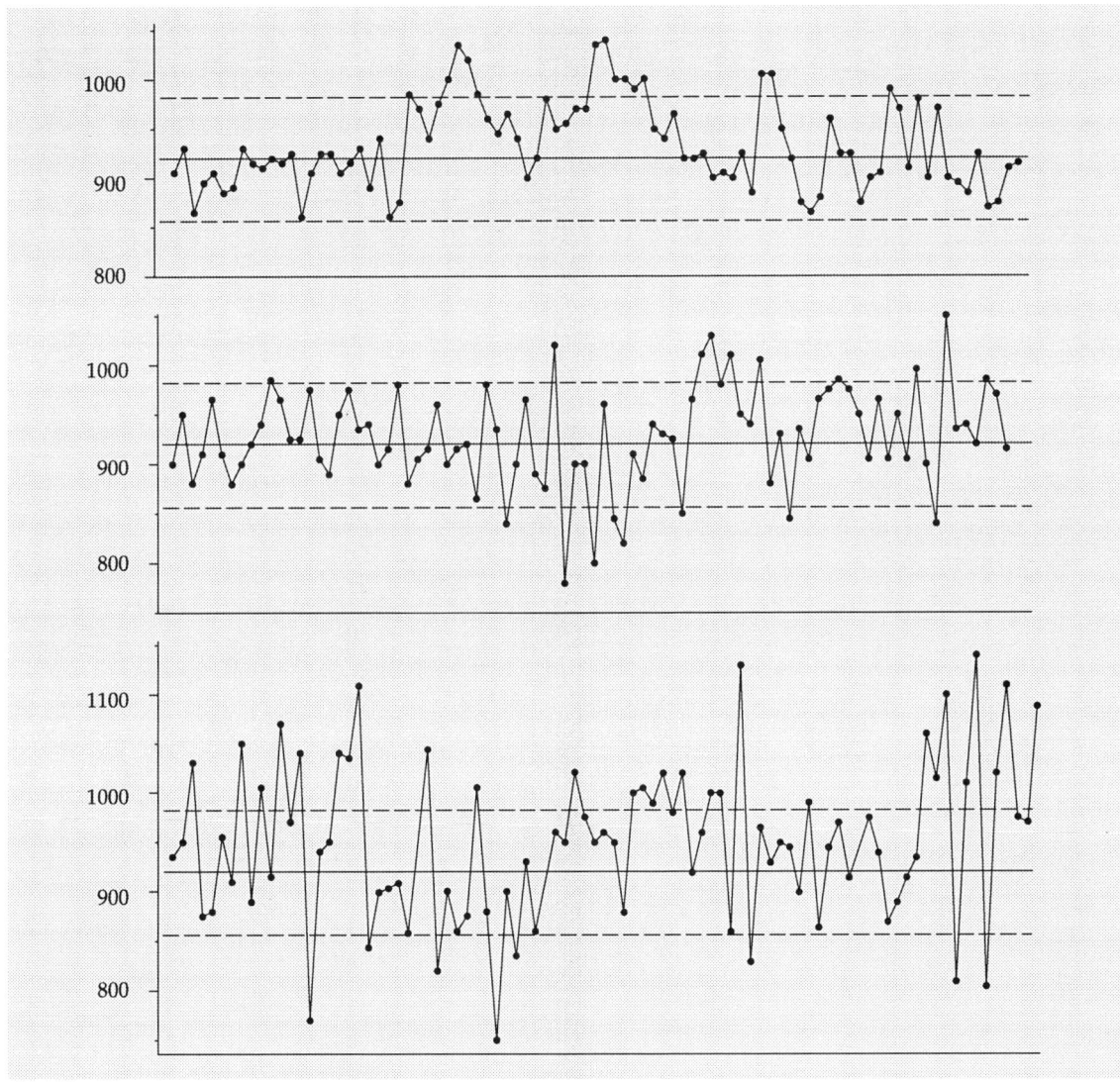

Figure 5.5: X Chart for the 259 Batch Weights

trace ingredients are added to the mixer based on the assumption that each batch will be exactly 920 kilograms. When a batch is not 920 kg, it will have either too much, or too little, of the trace ingredients. So, as the amount of the bulk ingredients vary, the formulation also varies, and the unpredictable batch weights are the visible manifestation of the inconsistent formulation of the compound. Thus, this process is producing compound with an inconsistent formulation. Both the weights and the formulation are in the State of Chaos— unpredictable and nonconforming. Since it is not possible to blend compound batches, this variation in formulation translates into variation in how the compound works in subsequent operations.

You cannot compute your way out of the problem shown in Example 5.1. No amount of data manipulation will ever remedy the problem of an unpredictable process. When a process is unpredictable, the past cannot be used as a reliable guide to the future. However, when a process is unpredictable, the excessive variation may be thought of a being due to one or more assignable causes. By identifying these assignable causes, and by taking appropriate action to remove the their effects from the process, you can turn your process into a predictable process, where the past does provide a reliable guide to the future.

But what about your customer's demand for capability indexes? You may well have to supply your customer with these values, but when your process is unpredictable these values will be unreliable. So you are going to have to educate your customer. (You might begin by giving him a copy of this book.) Of course, when you have to admit to your customer that you have an unpredictable process, you had better be working on that process to make it predictable. While you may be uncomfortable with the thought of telling your customer that you have an unpredictable process, you are not likely to surprise him, remember he has to use your stuff—he already knows that it is inconsistent. It will generally be better to confess to an unpredictable process than to wait until your customer comes to you with irrefutable evidence that your process is unpredictable.

Example 5.2: The Yield Data

A sample is taken from a continuous process five times each day. One of the values obtained from each of these samples is a complex number known as the "yield." Since, for the purpose of this example, the exact nature of these values is not important, they are presented here without elaboration.

The specifications for this yield value are 3410 to 3550. Since the customer has asked for daily capability indexes, the manufacturer uses the average and the range of the five values from each day to compute these indexes. The yield values for August 1 were:

$$3534 \quad 3542 \quad 3532 \quad 3537 \quad 3532$$

The average for August 1 was 3535.4, while the range was 10. Dividing this range by 2.326 we get a value for *Sigma(X)* of 4.30. This results in Natural Process Limits of 3522.5 to 3548.3. The difference between these Natural Process Limits is 25.8 units. Dividing the space available of 140 units by the space required of 25.8 units gives a Capability Ratio of:

$$C_p = \frac{140}{25.8} = 5.43$$

The Distance to Nearer Spec is 3550 − 3535.4 = 14.6 units, which is the effective space available. The space required for half of this process is 12.9 units, thus the Centered Capability Ratio is:

$$C_{pk} = \frac{14.6}{12.9} = 1.13$$

Continuing in this manner for the next thirteen days of production yields:

Date	Yields					Average	Range	C_p	C_{pk}
8/1	3534	3542	3532	3537	3532	3535.4	10	5.43	1.13
8/2	3533	3524	3524	3525	3527	3526.4	9	6.03	2.02
8/3	3531	3526	3529	3524	3527	3527.4	7	7.75	2.50
8/4	3525	3522	3521	—	—	3522.7	4	9.88	3.86
8/5	3521	3521	3521	3521	3515	3519.8	6	9.05	3.90
8/6	3498	3498	3506	3513	3536	3510.2	38	1.43	0.81
8/7	3526	3529	3524	3525	3520	3524.8	9	6.03	2.17
8/8	3517	3517	3519	3516	3517	3517.2	3	18.09	8.48
8/9	3453	3445	3451	3445	3452	3449.2	8	6.78	3.80
8/10	3445	3449	3454	3447	3446	3448.2	9	6.03	3.29
8/11	3440	3423	3416	3419	3415	3422.6	25	2.17	0.39
8/12	3458	3457	3457	3452	3446	3454.0	12	4.52	2.84
8/13	3448	3451	3453	3453	3455	3452.0	7	7.75	4.65
8/14	3475	3475	3474	3486	3490	3480.0	16	3.39	3.39

The histogram for these 68 individual values is shown in Figure 5.6. All of these individual values fall inside the specification limits of 3410 to 3550—so we can say that we have 100 percent conforming product.

The C_p values are spectacular—they average between 6.0 and 7.0, going as high as 18, with only one value below 2.0! So we can say that we have plenty of elbow room within these specifications.

The C_{pk} values are almost as incredible as the C_p values—they average about 3.0, going as high as 8.5, with only two values below 1.00. Clearly, with numbers this great, this process has got to be one of the best processes you have ever seen!

Yet there is more to the story than just the numbers. The 68 individual values used to create these phenomenal capability indexes are shown on an XmR chart in Figure 5.7. There the spikes on the moving range chart identify six distinct breaks in the production. This process is clearly unpredictable. It is subject to sudden, large changes in level, and in between these upsets it shows additional evidence of drifting around. While they have managed to stay within the specifications during this two-week period, there is no way to assure the customer that they will be within the specifications tomorrow, much less next week, or next month. This process is currently on the Brink of Chaos, but based on the suddenness of the changes shown in Figure 5.7, this process could jump into the State of Chaos at any time, without any warning.

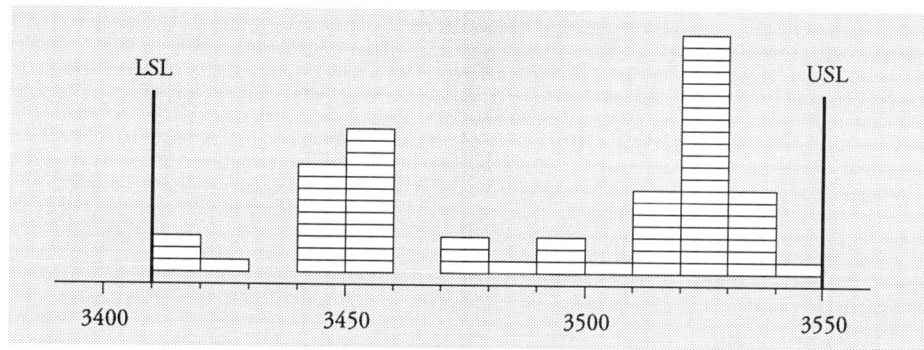

Figure 5.6: Histogram of 68 Yield Values

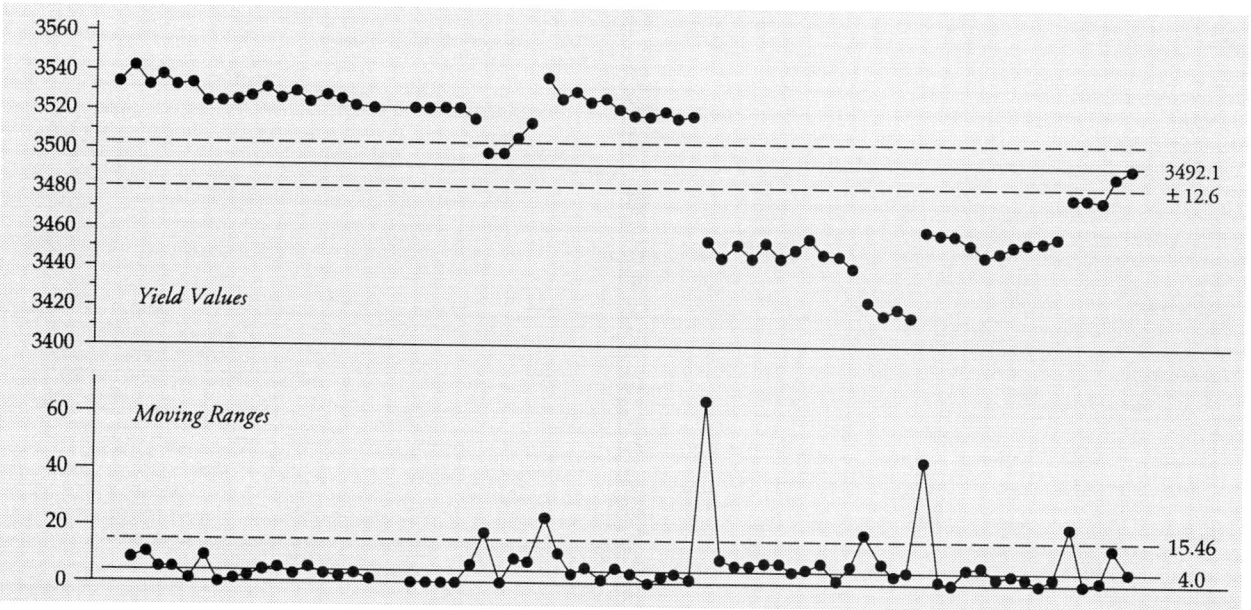

Figure 5.7: XmR Chart of the Yield Values

So what can you say about the capability of an unpredictable process? Very little. And even that little bit is merely wishful thinking.

"But I have an unpredictable process. What can I do to assure my customer that I can participate in a lean production scheme? How can I assure my customer that I can ship him conforming product without having to resort to inspection and sorting?" The honest answer is that you can't—there is no reliable way to characterize the future output of an unpredictable process. And if your process is not predictable, then you cannot truly participate in a lean production system.

5.2 What About Performance Indexes?

Since the term capability involves the notion of potential, and since an unpredictable process is not operating up to its potential, some have suggested computing *performance indexes* for unpredictable processes. The idea is that while we may not be able to characterize the future of an unpredictable process, there will still be a past history which we can describe in some manner—the process may not be predictable, but it will still have a track record. Of course, as anyone who has ever bet on a horse race can tell you, interpreting a track record is an uncertain science at best.

The Performance Ratio, P_p, like the Capability Ratio, has a numerator that defines the space available for the product: [upper specification limit – lower specification limit]. However, the denominator for the Performance Ratio is different from that of the Capability Ratio. While the Capability Ratio uses a "within-subgroup" measure of dispersion which comes from a process behavior chart in the denominator, the Performance Ratio uses an "overall" measure of dispersion in the denominator. This overall (global) measure of dispersion is computed using all of the data collected over a given period of time. Since we are discussing an unpredictable process, these two different ways of measuring dispersion will give considerably different results. The Performance Ratio is defined to be

$$\text{Performance Ratio} = P_p = \frac{\text{USL} - \text{LSL}}{6s}$$

where s is the standard deviation statistic computed by using all of the data together in a single computation.

The Centered Performance Ratio, P_{pk}, is obtained by dividing the DNS value by $3s$. Again, P_{pk} is similar to the Centered Capability Ratio, except that it uses a global measure of dispersion in the denominator.

$$\text{Centered Performance Ratio} = P_{pk} = \frac{\text{DNS}}{3s}$$

<u>Example 5.3:</u> <u>The Batch Weight Data</u>

The Batch Weight Data for the first shipment consisted of the thirty values shown below.

```
905 930 865 895 905    885 890 930 915 910    920 915 925 860 905
925 925 905 915 930    890 940 860 875 985    970 940 975 1000 1035
```

The average weight of these 30 batches is 920.8 kg. The standard deviation statistic computed using these 30 values is $s = 40.47$ kg. The specifications for these weights were taken to be 850 kg to 990 kg, thus, the Elbow Room is $990 - 850 = 140$ kg. Dividing this value by $6s = 242.8$ kg gives a Performance Ratio of:

$$P_p = \frac{140 \text{ kg}}{242.8 \text{ kg}} = 0.58$$

The Distance to the Nearer Spec for these 30 values is $990 - 920.8 = 69.2$ kg. To obtain a Centered Performance Ratio we divide this value by $3s = 121.4$ kg to get:

$$P_{pk} = \frac{69.2 \text{ kg}}{121.4 \text{ kg}} = 0.57$$

These performance indexes are considerably different from the capability indexes of 1.11 and 1.10 computed in Example 5.1. The customer will probably interpret these performance indexes to mean that the space available is about 57 percent of the space required. But what does this mean? If we assume a normal distribution, these performance indexes would convert into about 8 percent nonconforming, which is similar to the 7 percent that actually occurred.

At this point we might be tempted to characterize this process as having about 8 percent nonconforming. But remember, the process is unpredictable. We have already seen, in Example 5.1, the futility of trying to use the past as a guide to the future when dealing with an unpredictable process.

So here we will look at the use of performance indexes as characterizations of past shipments. Recall that in this case a shipment consists of 30 batches. The following table summarizes each of the shipments by showing the average, the standard deviation, the performance indexes, the associated theoretical percent nonconforming from Table 4.1, and the actual percent nonconforming.

Batches	Average	s	P_p	P_{pk}	Theoretical % Nonconforming	Actual % Nonconforming		
1–30	920.8	40.47	0.58	0.57	8 %	2/30	or	7 %
31–60	955.3	41.48	0.56	0.28	21 %	6/30	or	20 %
61–90	920.5	41.57	0.56	0.56	9 %	2/30	or	7 %
91–120	927.7	33.42	0.70	0.62	4 %	0/30	or	0 %
121–150	919.0	65.62	0.36	0.35	28 %	10/30	or	33 %
151–180	938.3	48.30	0.48	0.36	18 %	5/30	or	17 %
181–210	928.8	91.03	0.26	0.22	44 %	14/30	or	47 %
211–240	960.7	58.23	0.40	0.17	33 %	9/30	or	30 %
241–259	976.1	97.4	0.24	0.05	54 %	10/19	or	52 %

While the performance indexes do serve to characterize each batch, they do so by computing a theoretical percentage nonconforming that merely mimics the actual percentage nonconforming. Compare the last two columns above. Once the process behavior chart (see Figure 5.5) has shown us that this process is unpredictable, about all that we are left with is a description of the past shipments. The actual fraction nonconforming provides this summary of the past shipments without a lot of computation. The performance indexes dress this same information up in jargon and fancy computations that, in the end, do not add anything to our knowledge beyond the fraction nonconforming. Because of the unpredictable nature of this process, we cannot use the capability indexes, nor the performance indexes, to extrapolate into the future. And as a description of the past they are more complex than they need to be.

Example 5.4: The Yield Data

We have already seen wildly variable capability indexes computed on the daily yield values. The same holds true if we compute daily performance indexes:

Date	Yields					Average	s	P_p	P_{pk}
8/1	3534	3542	3532	3537	3532	3535.4	4.22	5.53	1.15
8/2	3533	3524	3524	3525	3527	3526.4	3.78	6.17	2.06
8/3	3531	3526	3529	3524	3527	3527.4	2.70	8.64	2.79
8/4	3525	3522	3521	—	—	3522.7	2.08	11.21	4.38
8/5	3521	3521	3521	3521	3515	3519.8	2.68	8.70	3.75
8/6	3498	3498	3506	3513	3536	3510.2	15.72	1.48	0.84
8/7	3526	3529	3524	3525	3520	3524.8	3.27	7.13	2.57
8/8	3517	3517	3519	3516	3517	3517.2	1.10	21.30	9.98
8/9	3453	3445	3451	3445	3452	3449.2	3.90	5.98	3.35
8/10	3445	3449	3454	3447	3446	3448.2	3.56	6.55	3.57
8/11	3440	3423	3416	3419	3415	3422.6	10.21	2.28	0.41
8/12	3458	3457	3457	3452	3446	3454.0	5.05	4.62	2.90
8/13	3448	3451	3453	3453	3455	3452.0	2.65	8.82	5.29
8/14	3475	3475	3474	3486	3490	3480.0	7.45	3.13	3.13

If we convert these performance indexes into fractions nonconforming we will get zero percent for 12 of the 14 dates. On August 6 we have a P_{pk} value of 0.84, which would correspond to a theoretical percentage nonconforming of 0.5%. On August 11 we have a P_{pk} value of 0.41, which would correspond to a theoretical percentage nonconforming of 11%. With these two exceptions, these performance indexes suggest that they have managed to keep this process within the specifications—which is a fair assessment of these data, all 68 of the values are within the specifications.

However, if we use all of the 68 Yield values collectively to compute a global pair of performance indexes we get a different story. The average of these 68 values is 3491.25 units and the standard deviation is $s = 38.52$ units. The specification limits were 3410 to 3550. Thus our global performance indexes would be:

$$P_p = \frac{140 \text{ units}}{231.1 \text{ units}} = 0.61 \quad \text{and} \quad P_{pk} = \frac{58.75 \text{ units}}{115.6 \text{ units}} = 0.51$$

If these values were converted they would correspond to about 8 percent nonconforming. However, as noted above, none of the values were outside the specifications. As unpredictable as this process was, they managed to stay inside the specification limits.

So which performance indexes would you use? The daily values look good, while the global values look bad. Which would you use if you were the supplier? Which would you use if you were the customer? Will performance indexes ever provide a consensus on how to characterize the performance of this process?

No assumption you can make, and no calculation that you can perform, can change the fact that an unpredictable process is inherently inconsistent. Try as you might, the unpredictability will undermine all attempts to characterize an unpredictable process.

So while you may compute performance indexes, the main benefit of doing so is to practice your arithmetic. A better way of characterizing the past for an unpredictable process is simply to report the fraction nonconforming, with a warning that the past is not a reliable guide to the future. You say you don't want to tell your customer this? If he is using process behavior charts on your material he already knows. This is not something that you can hide from those that know how to properly separate routine variation from exceptional variation.

5.3 Hypothetical Capability

While your process may be unpredictable, there is still a way to approximate what your process has the *potential* to do. This potential will remain purely hypothetical unless and until you learn how to operate and maintain your process in a predictable state. Nevertheless, there are times when it is helpful to know where you are likely to end up before you set out on a journey. Likewise, it may be useful to know what your process has the potential to do even while it is operating unpredictably.

This hypothetical potential of an unpredictable process is defined by the process behavior chart. In fact, that is how the process behavior chart works—it defines the potential for the process and then compares the behavior of the data against this potential. If the process behavior is consistent with this potential, then the process is said to be predictable, otherwise it is unpredictable.

Thus, the Natural Process Limits obtained from a process behavior chart define (1) the actual capability of a predictable process, and (2) the potential capability of an unpredictable process. Since the only time that a process will actually operate up to its potential is when it is operated predictably, this capability will remain hypothetical unless you take action to find and remove causes of exceptional variation whenever they occur.

Example 5.5: The Batch Weight Data

The *XmR* chart in Figure 5.3 shows an unpredictable process. However, the median moving range of 20 kg is still an estimate of the inherent batch-to-batch variation of this process. The Natural Process Limits shown on the *X* chart were computed from this median moving range, and therefore they provide a reasonable approximation of the consistency that this process might deliver if they make the effort to find and remove the assignable causes that dominate this process. If they can also get this process to operate with an average that is near the target of 920 kg, then they are likely to have a Capability Ratio in the neighborhood of 1.11. Thus, the numbers calculated from the process behavior chart provide a rough approximation of what to expect if you take the trouble to operate and maintain your process in a predictable state.

Example 5.6: The Yield Data

With the yield data a set of capability indexes was computed every day. These daily values varied dramatically. This variation was the result of two things: the small amount of data used for each computation, and the wildly unpredictable nature of this process. However, the moving range chart still characterizes the short-term variation of this process. The six moving ranges that fall above the limit identify the six abrupt changes in this process. Except for these six upsets, the moving range chart shows a consistent amount of variation for the remainder of this period. The Natural Process Limits shown in Figure 5.7 were based on the median moving range value of 4. These limits suggest that this process is capable of being operated with about ±12.6 units of variation. If they could simply maintain a consistent level, they could get virtually all of their values to fall within a range of 25 units. A median moving range of 4.0 gives a *Sigma(X)* value of:

$$Sigma(X) = \frac{4.0}{0.954} = 4.2 \text{ units}$$

This suggests that the specification limits of 3410 to 3550 are about 33 standard deviations wide, and when predictable, this process could have capability indexes in the 5.0 to 6.0 range. With this much elbow room, a predictable process could be operated high or low, wherever it is most economical, and still have a comfortable margin between the Natural Process Limits and the nearer specification.

Beyond Capability Confusion

While both of the processes considered in this chapter have the potential of being capable, neither process is currently meeting that potential. Nor will they ever meet that potential until they are operated predictably. It doesn't matter how good the capability indexes may be, as long as the process is unpredictable, its performance will always fall short of its potential. And process potential is a terrible thing to waste.

5.4 A Flowchart for Process Capability

Before a process can be said to have a well-defined capability, it must be predictable. Therefore, the *capability* of a process depends upon the *predictability* of the process. When a process is predictable virtually all of the individual values will fall within the Natural Process Limits, and these limits *define* the capability of the process.

Figure 5.8 shows a flowchart for assessing the capability of a process. It begins with a process capability chart since this chart is the only way to answer the question of whether or not the process is predictable.

If the process is unpredictable, then there is no well-defined capability, as was shown in the Section 5.1. You may consider what the hypothetical capability may be, as outlined in Section 5.2, but this potential will not be realized until you take the time and make the effort to operate and maintain the process in a predictable manner. Therefore, the first exit from the flowchart is labeled "work on operating process predictably."

If the process has behaved predictably in the past, then it may be considered to have a well-defined capability and this capability may be most easily shown by plotting a histogram of the individual values from the process behavior chart and plotting the specification limits on this histogram. The relationship between the histogram and these limits will compare the Voice of the Customer and the Voice of the Process.

Next the capability indexes may be computed. The Specified Tolerance expressed in sigma units and divided by 6.0 will give the Capability Ratio, C_p. If the Specified Tolerance is greater than six standard deviations, or C_p is greater than 1.0, then the process has the Elbow Room to operate with virtually 100 percent conforming product.

The Distance to Nearer Spec expressed in sigma units and divided by 3.0 will give the Centered Capability Ratio, C_{pk}. If the DNS value is greater than three standard deviations, or C_{pk} is greater than 1.0, then the process should be operating with virtually 100 percent conforming product.

If the Distance to Nearer Spec is less than three standard deviations then there process may be located too close to the specification limit. Refer to past history, the histogram, and your knowledge about this process to figure out whether or not the process is capable. If it is not capable, then you will need to work on setting the process aim correctly.

If the Specified Tolerance is less than six standard deviations then there may not be enough elbow room for the process. Refer to past history, the histogram, and your knowledge about this process to figure out whether or not the process is capable. If it is not capable, you will need to set the process aim to minimize nonconforming product, and work on reducing the process variation or loosening the product specifications.

Thus, there are four different outcomes for the flowchart in Figure 5.8. The first of the four outcomes corresponds to a process which is either in the State of Chaos or the Brink of Chaos. When your process is in either of these two states you will need to work on finding and removing the assignable causes that affect your process. Unless and until you do this your process will continue to be unpredictable, and this

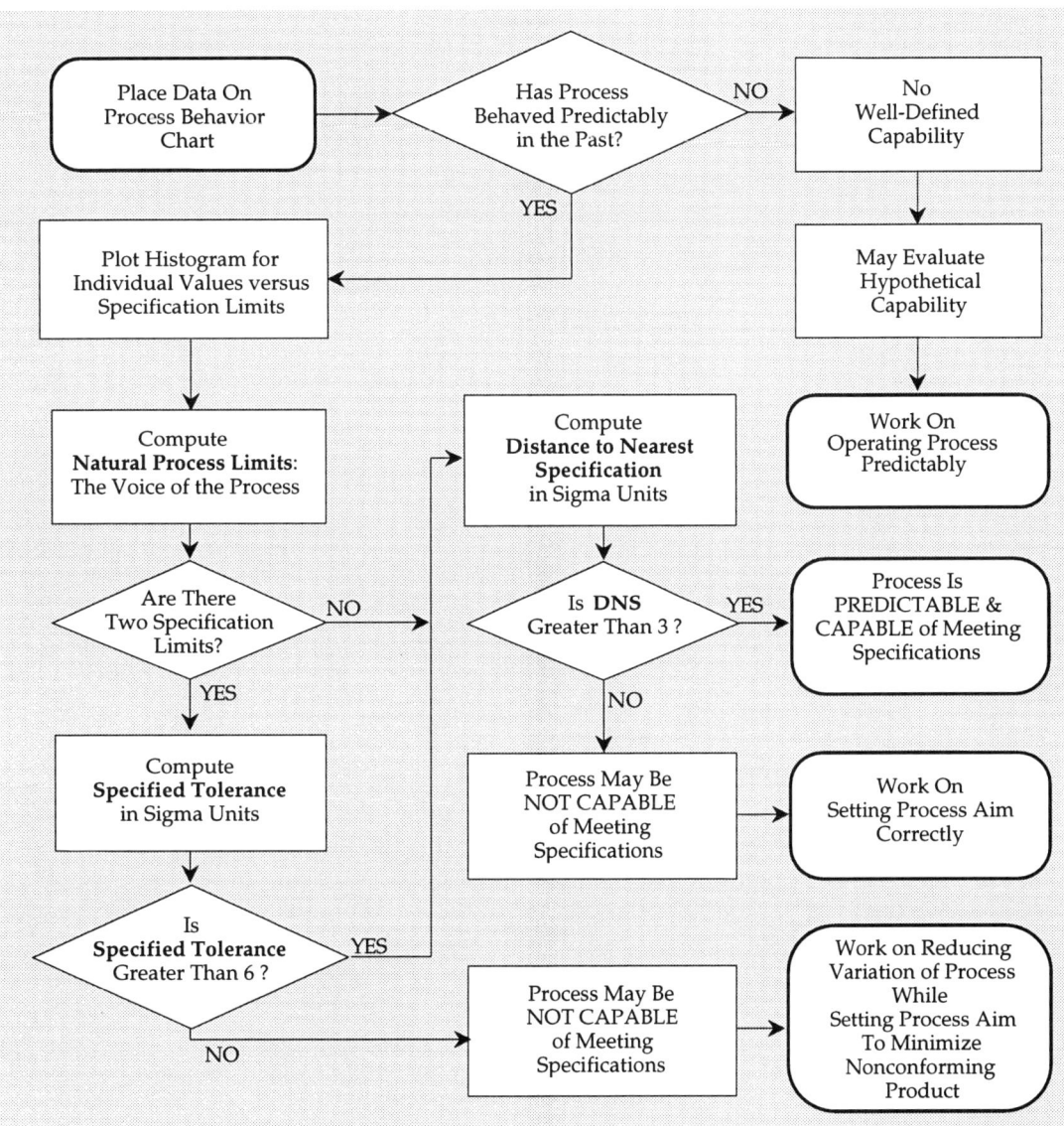

Figure 5.8: Assessing Process Capability

unpredictable variation will end up reducing your productivity while it undermines your product quality.

The second of the four outcomes corresponds to a process which is in the Ideal State. Here you may expect your process to continue to make 100 percent conforming product as long as you continue to use the process behavior chart to maintain and operate the process predictably.

The last two outcomes shown may correspond to processes which are either in the Threshold State or in the Ideal State. Here the predictability of your process will allow you to extrapolate from the past to the future with some reasonable degree of assurance. In addition, the continued use of the process behavior chart will allow you to identify when opportunities for improvement occur.

5.5 Summary

When your process is unpredictable the past is not a reliable guide to the future.

Performance indexes are descriptive measures of the past that do not provide any useful information beyond an estimate of the fraction nonconforming.

The best estimate of the fraction nonconforming will always be computed directly from the data.

However, the Natural Process Limits obtained from a process behavior chart will provide an approximation of the hypothetical capability of an unpredictable process. Of course, the question is what are you going to do to turn this hypothetical capability into reality?

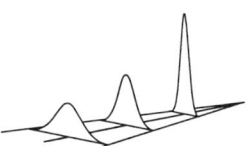

Chapter Six

Capability Ratios Vary!

All statistics will vary from sample to sample even when the underlying process does not change. Capability Ratios are statistics. To see how capability ratios vary from sample to sample we will use the following data.

Table 6.1: 100 Subgroups of Size 5 Drawn from a Predictable Process

Values					$\bar{X}$	R	Values					$\bar{X}$	R	Values					$\bar{X}$	R
12	9	9	8	9	9.4	4	9	10	10	11	10	10.0	2	10	12	11	9	13	11.0	4
11	13	11	11	11	11.4	2	11	8	11	9	11	10.0	3	10	7	11	8	13	9.8	6
14	6	11	8	9	9.6	8	12	11	13	8	10	10.8	5	12	10	9	8	13	10.4	5
10	12	13	8	10	10.6	5	9	11	11	11	11	10.6	2	12	10	11	10	10	10.6	2
11	10	12	11	8	10.4	4	13	11	10	7	7	9.6	6	11	10	12	10	9	10.4	3
7	10	7	7	9	8.0	3	10	9	13	10	12	10.8	4	7	8	12	10	10	9.4	5
12	12	14	7	10	11.0	7	8	9	7	10	11	9.0	4	8	9	11	9	8	9.0	3
11	10	9	10	10	10.0	2	7	15	7	8	11	9.6	8	9	14	12	12	11	11.6	5
9	9	11	11	9	9.8	2	11	9	10	11	13	10.8	4	7	8	9	9	9	8.4	2
11	12	10	9	11	10.6	3	10	13	10	9	11	10.6	4	12	9	10	10	9	10.0	3
9	10	11	11	10	10.2	2	12	9	7	13	13	10.8	6	8	9	9	11	7	8.8	4
9	9	10	11	7	9.2	4	9	11	9	10	9	9.6	2	10	8	12	9	11	10.0	4
7	10	12	11	10	10.0	5	8	15	11	10	9	10.6	7	12	10	10	10	13	11.0	3
8	10	11	9	10	9.6	3	7	9	5	12	13	9.2	8	9	9	10	8	9	9.0	2
10	8	11	12	11	10.4	4	10	9	10	11	10	10.0	2	10	9	7	9	9	8.8	3
11	8	9	11	7	9.2	4	12	12	10	11	11	11.2	2	11	7	5	9	8	8.0	6
9	9	11	11	8	9.6	3	11	11	12	10	14	11.6	4	9	11	11	10	9	10.0	2
12	9	11	10	6	9.6	6	12	9	12	9	9	10.2	3	12	10	13	12	13	12.0	3
12	11	10	15	12	12.0	5	8	10	14	13	9	10.8	6	12	10	11	8	12	10.6	4
9	12	11	9	9	10.0	3	10	10	10	14	11	11.0	4	10	9	12	9	13	10.6	4
13	11	12	7	8	10.2	6	11	11	8	11	7	9.6	4	11	12	9	11	7	10.0	5
9	9	13	8	7	9.2	6	10	9	11	8	10	9.6	3	7	8	13	10	10	9.6	6
10	12	9	11	10	10.4	3	11	11	10	10	13	11.0	3	9	9	8	7	9	8.4	2
8	8	12	10	8	9.2	4	12	11	5	9	7	8.8	7	9	10	11	9	7	9.2	4
9	13	7	10	13	10.4	6	11	9	9	11	12	10.4	3	10	11	11	9	10	10.2	2
11	10	12	10	10	10.6	2	14	11	14	10	8	11.4	6	12	9	13	9	8	10.2	5
11	9	9	8	8	9.0	3	9	10	10	15	11	11.0	6	11	13	11	11	7	10.6	6
11	12	8	12	10	10.6	4	10	13	7	12	10	10.4	6	7	8	9	10	8	8.4	3
11	14	8	13	8	10.8	6	11	9	12	10	13	11.0	4	9	11	12	11	9	10.4	3
14	12	9	9	10	10.8	5	6	10	11	10	10	9.4	5	9	10	9	10	10	9.6	1
9	11	13	10	7	10.0	6	10	10	11	7	10	9.6	4							
10	11	10	12	11	10.8	2	10	12	12	8	10	10.4	4							
9	10	9	13	14	11.0	5	13	7	11	12	11	10.8	6							
12	10	9	8	8	9.4	4	12	10	7	10	9	9.6	5							
9	7	14	12	9	10.2	7	9	9	9	10	10	9.4	1							

Example 6.1:   The Quincunx Data

The data in Table 6.1 were obtained using a "quincunx" (also known as a bead board). Since no changes were made in the settings of the bead board while these 500 data were collected, these data should be taken as being typical of data from a process which displays a reasonable degree of predictability. And that is exactly what we see in Figure 6.1—an average and range chart that reveals a predictable process—there are no indications of anything other than routine variation.

Beyond Capability Confusion

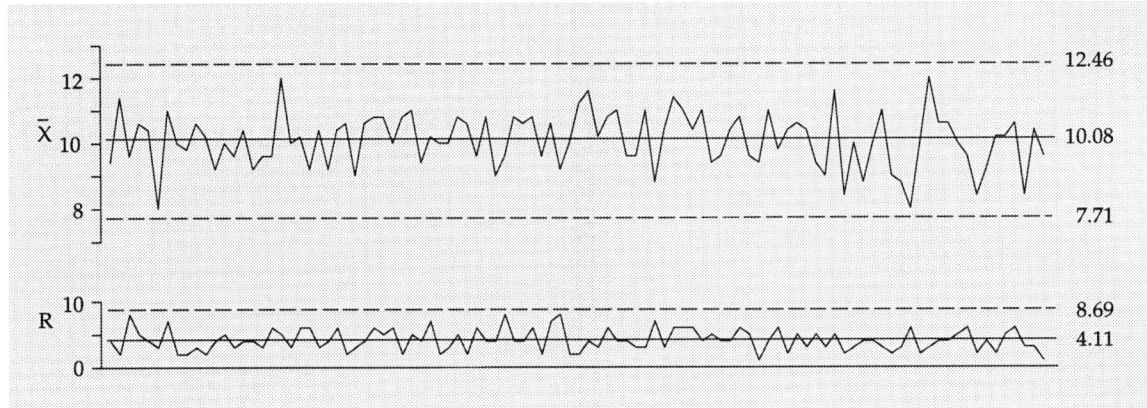

Figure 6.1: Average and Range Chart for the Data of Table 6.1.

For the chart in Figure 6.1 the Grand Average is 10.084 and the Average Range is 4.11. Thus, the Natural Process Limits would be computed as:

$$\bar{\bar{X}} \pm 3\frac{\bar{R}}{d_2} = 10.084 \pm 3\,(1.767) = 4.78 \text{ to } 15.39$$

meaning that we should expect process outcomes that vary from 5 to 15. And that is what we see when the 500 values are plotted in a histogram. If we set the specification limits for this process at 3 and 17, then we will have a process in the ideal state—both predictable and capable.

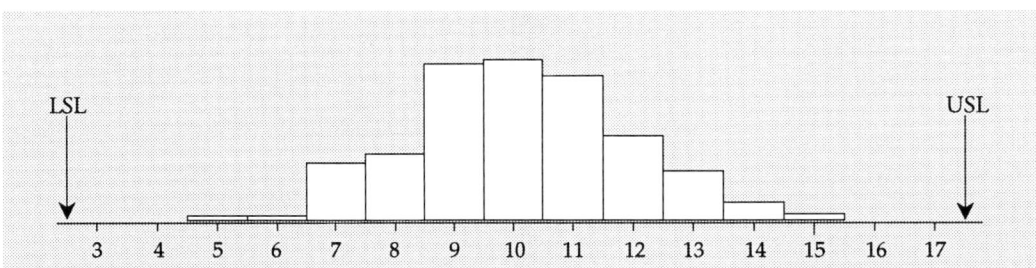

Figure 6.2: Histogram and Specifications for the Data of Table 6.1

With these specification limits, and based on the summary statistics from the process behavior chart, the Capability Ratio for this bead board process would be:

$$C_p = \frac{\text{USL} - \text{LSL}}{6\,Sigma(X)} = \frac{(17-3)}{(6.0)\,(1.767)} = 1.32.$$

With a Grand Average of 10.084, the Centered Capability Ratio is:

$$C_{pk} = \frac{\text{USL} - \bar{\bar{X}}}{3\,Sigma(X)} = \frac{17 - 10.084}{(3.0)\,(1.767)} = 1.30.$$

Today many customers ask for Centered Capability Ratios of 1.33 or greater. While this process is close, it is not quite meeting this magic value.

6.1 Capability Indexes Over Time

The calculations in Example 6.1 were based on 500 data. You will seldom wait until you have so many data to compute a capability index. In most cases capability indexes are computed using 50 values or less. To see the effect that this has on the capability indexes consider the following examples.

Example 6.2:   The Quincunx Data

Instead of using all of the data from Table 6.1, consider what would happen if the Centered Capability Ratio was computed every ten subgroups:

Subgroups	Grand Average	Average Range	C_{pk}	What your boss had to say about this C_{pk} value:
1 to 10	10.08	4.0	1.34	*A good capability ratio, our customer will be pleased with this.*
11 to 20	9.98	3.9	1.39	*Congratulations, the capability ratio got better.*
21 to 30	10.12	4.5	1.19	*What happened to cause the capability ratio to drop?*
31 to 40	10.24	4.2	1.25	*This capability ratio is still not good enough—we must do better.*
41 to 50	10.10	4.9	1.09	*This is the third value below 1.33—things had better improve or else!*
51 to 60	10.42	3.9	1.31	*Well, this is better, but our customer is asking for 1.33 or greater.*
61 to 70	10.30	4.7	1.11	*You were supposed to be improving—what is going on here?*
71 to 80	10.06	3.8	1.42	*At last, send these data to our customer.*
81 to 90	9.88	3.5	1.52	*Excellent capability ratio. Good work.*
91 to 100	9.66	3.7	1.40	*Why is this value slipping?*

Of course, the real answer is that we are not slipping. There is no evidence that the process has changed in any way during this period of time. The process was not getting worse, nor was it getting better. But the statistics were changing with each set of subgroups even though the process remained the same, and these changes resulted in different values for the Centered Capability Ratio.

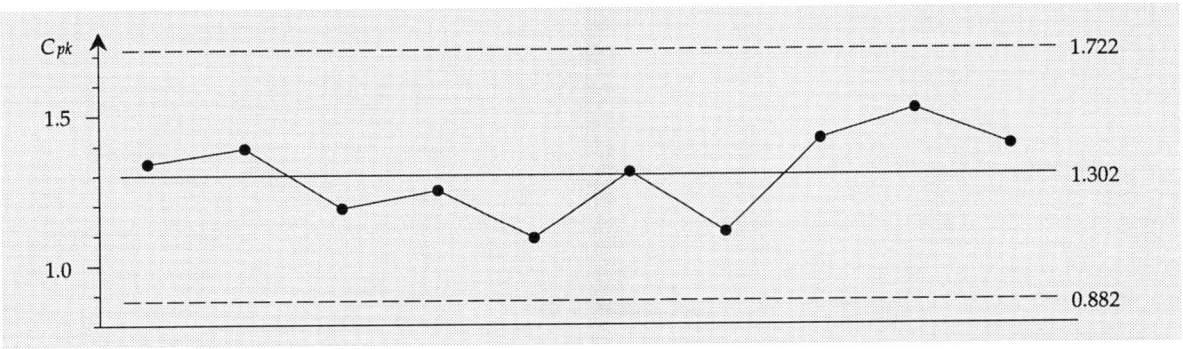

Figure 6.3: Individual Chart for Centered Capability Ratios Based on 50 Values Each

Figure 6.2 shows that the routine amount of variation in Centered Capability Ratios, each computed using 50 values, will allow them to go as low as 0.88, or as high as 1.72, while the process is unchanged with an average capability ratio of 1.30. *Before you can interpret a single capability ratio, based upon 50 data, to mean that this process capability has changed from the value of 1.30, the single ratio would have to fall below 0.88, or it would have to go above 1.72.* Think about what this means about how you interpret capability indexes.

Example 6.3:   The Percent Solids Data

In Example 4.5 the percent solids values for a continuous process were used to characterize the process as predictable but not capable. The effective lower specification limit was 45.5. Instead of using all 84 values to compute a Centered Capability Ratio they might want to use the six values obtained each day to compute a daily ratio (for subgroups of size six the d_2 value is 2.534):

Day	Values						Average	Range	C_{pk}
1	48	50	49	48	50	50	49.17	2	1.55
2	45	51	47	49	53	51	49.33	8	0.41
3	50	50	47	46	48	47	48.00	4	0.53
4	45	46	47	50	47	49	47.33	5	0.31
5	50	45	50	47	50	49	48.50	5	0.51
6	46	48	46	46	51	48	47.50	5	0.34
7	49	51	48	43	46	48	47.50	8	0.21
8	47	49	49	51	46	47	48.17	5	0.45
9	47	46	50	50	48	49	48.33	4	0.60
10	49	46	48	47	50	48	48.00	4	0.53
11	48	49	47	47	49	48	48.00	2	1.06
12	48	49	47	47	49	48	47.83	6	0.33
13	46	44	48	46	49	50	47.17	6	0.24
14	47	48	47	48	47	47	47.33	1	1.55

Each of these daily Centered Capability Ratios is based on a very small amount of data, and therefore they vary more than did the ratios based on 50 data in the previous example. While you may not like the thought of using only six data to compute a capability index, and while it may not be recommended, it is a fairly common practice. If we place these 14 values on an *XmR* Chart we get Figure 6.4.

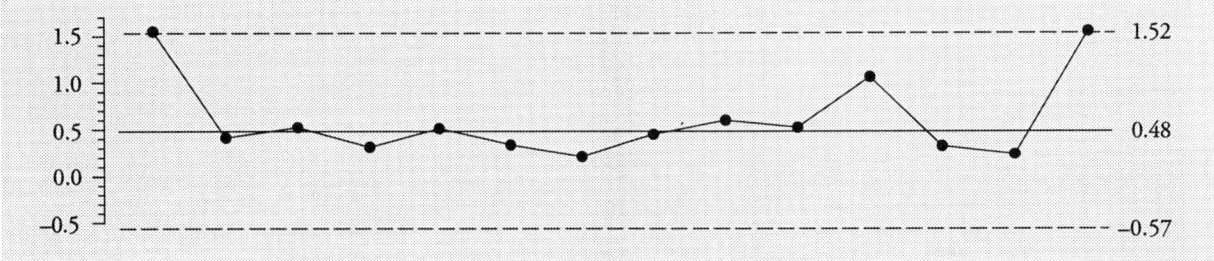

Figure 6.4: *X* Chart for the Daily Centered Capability Ratios for the Percent Solids Data

The central line in Figure 6.4 is the median of the 14 values, rather than their average, because the two large values unduly inflate the average. The small amount of data used to compute each ratio results in increased variation and wider limits than we saw in Figure 6.3. While the process is predictable but not capable, the daily Centered Capability Ratios can vary from –0.57 to 1.52 without indicating any change in the process capability.

The central line on the chart in Figure 6.4 is 0.48, while the C_{pk} value found in Example 4.5 was 0.45. Therefore, it would appear that the effective space available for this process is around 45 percent to 48 percent of the generic space required. Table 4.1 suggests that we should expect about 8 or 9 percent defective from this process unless we take steps to change things.

But what do we make of the two points above the limits in Figure 6.4? Since the process data reveal the process itself to be predictable, we should interpret the two points outside the limits in Figure 6.4 to mean that these two samples are poor samples—the data from these two days do not provide an appropriate cross-section for characterizing the full extent of the process variation.

The *XmR* chart for the capability indexes computed for a single process at different points in time will graphically show you the inherent uncertainty that is contained in your computations. When you have capability indexes based on large amounts of data from a predictable process the indexes should be fairly steady, with minimal uncertainty, resulting in an *X* Chart with reasonably tight limits. When you have capability indexes based on small amounts of data they will be full of uncertainty, resulting in wide limits. In any case, the central line will tend to be a better estimate of the actual status of the predictable process than will the individual capability indexes plotted on the chart.

But what happens if we use this same approach with capability indexes computed for an unpredictable process? (By now you should know better than to calculate capability indexes for an unpredictable process. Or, if you are forced to do so, you should know better than to believe such indexes. Nevertheless, since there is a lesson to be learned here, bear with me while I compute capability indexes for unpredictable processes…)

Example 6.4: The Yield Data

In Example 5.2 the yield data were used to characterize a process as unpredictable but conforming. There the capability indexes were computed on a daily basis using five values each. If we place the Centered Capability Ratios on an *XmR* chart we end up with Figure 6.5.

Date	Yields					Average	Range	C_p	C_{pk}
8/1	3534	3542	3532	3537	3532	3535.4	10	5.43	1.13
8/2	3533	3524	3524	3525	3527	3526.4	9	6.03	2.02
8/3	3531	3526	3529	3524	3527	3527.4	7	7.75	2.50
8/4	3525	3522	3521	—	—	3522.7	4	9.88	3.86
8/5	3521	3521	3521	3521	3515	3519.8	6	9.05	3.90
8/6	3498	3498	3506	3513	3536	3510.2	38	1.43	0.81
8/7	3526	3529	3524	3525	3520	3524.8	9	6.03	2.17
8/8	3517	3517	3519	3516	3517	3517.2	3	18.09	8.48
8/9	3453	3445	3451	3445	3452	3449.2	8	6.78	3.80
8/10	3445	3449	3454	3447	3446	3448.2	9	6.03	3.29
8/11	3440	3423	3416	3419	3415	3422.6	25	2.17	0.39
8/12	3458	3457	3457	3452	3446	3454.0	12	4.52	2.84
8/13	3448	3451	3453	3453	3455	3452.0	7	7.75	4.65
8/14	3475	3475	3474	3486	3490	3480.0	16	3.39	3.39

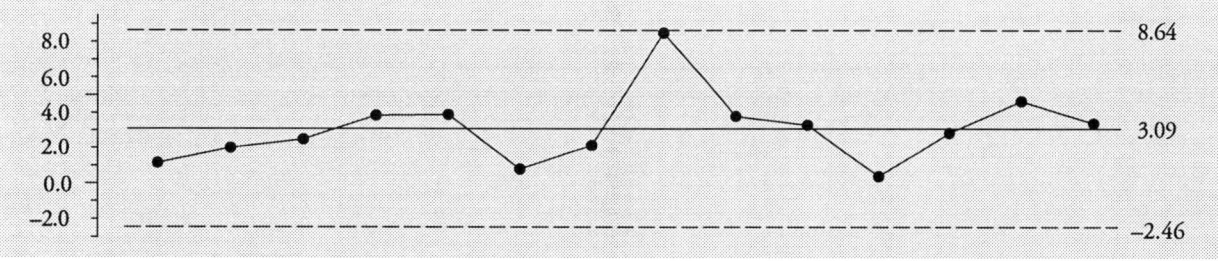

Figure 6.5: *X* Chart for the Daily Centered Capability Ratios for the Yield Data

The Centered Capability Ratios in Figure 6.5 show a huge amount of variation. This variation comes from two sources: the small amounts of data used to compute each ratio, and the unpredictability of the underlying process. Instead of getting excited when the capability ratio changes from day to day, they need to learn how to operate this process predictably.

Beyond Capability Confusion

Example 6.5: The Batch Weight Data

In Example 5.1 we used the batch weight data to characterize the process as unpredictable and nonconforming. If we computed the Centered Capability Ratio for each group of 30 batch weights we would get the following values:

Batches	Average	Median mR	C_{pk}
1–30	920.8	20	1.10
31–60	955.3	20	0.55
61–90	920.5	37.5	0.59
91–120	927.7	35	0.57
121–150	919.0	60	0.37
151–180	938.3	50	0.33
181–210	928.8	92.5	0.21
211–240	960.7	40	0.23
241–259	976.1	85	0.05

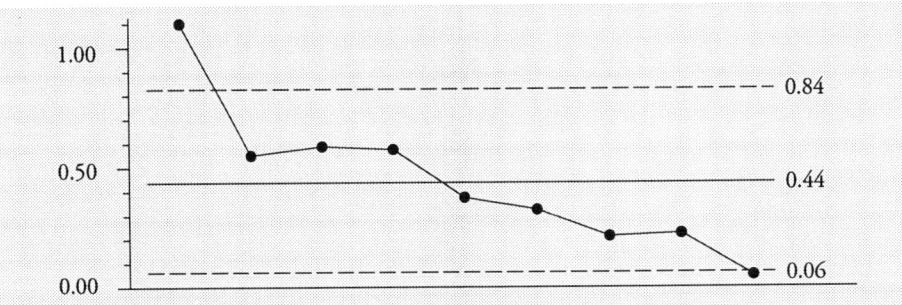

Figure 6.6: X Chart for the Shipment-by-Shipment Centered Capability Ratios for the Batch Weight Data

A glance back at Figure 5.5 shows that this process is deteriorating, and this deterioration may be seen in the downward trend in the Centered Capability Ratios in Figure 6.6. While the process is wildly unpredictable, with 68 of the 259 points outside the limits used, only two of the capability ratios fall outside the limits in Figure 6.6. Notice, however, that the limits are tighter here than they were in the two previous examples. This is due to the larger amount of data used in computing each of the capability indexes.

As the amount of data used to compute a given index increases the uncertainty in that index will decrease. This effect may be seen in the four preceding examples. However this effect is not a linear effect. Indexes computed with fewer than 15 values will tend to be quite variable, resulting in wide limits on an XmR Chart. Indexes computed using 40 or more values will tend to have stabilized, resulting in narrower limits on an XmR Chart. Whatever the amount of data you have used to compute your capability indexes, the XmR Chart will explicitly show the uncertainty in the indexes by the width of the limits. This will allow you to avoid the mistake made by the boss in Example 6.2—the mistake of overinterpreting the changes in the capability indexes.

Placing your Capability indexes on an XmR Chart is the best way to know if your *predictable process* is meeting a "capability target." If the central line of this X Chart is greater than the target, then you are meeting the target as long as the current capability index is within the limits. For example, the producer in Example 6.2 could certify that his process capability is 1.30, even though four of the ten values were less than 1.30.

Likewise, the producer in Example 6.3 could certify that his process capability was about 0.48.

6.2 Charting Capability Indexes Is No Substitute for Charting the Process

While you may place the capability ratios on a process behavior chart, that chart will not provide an effective way of monitoring the behavior of the underlying *process*.

In the first place, capability indexes are complex transformations of the original data. Since they are ratios they behave in a complex manner. This means that the process may behave one way while the computed capability indexes behave another way. In Example 6.2 the process was predictable, and so was the X Chart for the Centered Capability Ratios. In Example 6.3 the process was predictable, yet the X Chart had two of the 14 ratios above the upper limit. In Example 6.4 the process is unpredictable, yet the X Chart for the Centered Capability Ratios is predictable. And in Example 6.5 the process is unpredictable, and so is the X chart for the Centered Capability Ratios. In short, there is no particular relationship between the predictability of the process and the predictability of the capability indexes.

	Process is Predictable	Process is Unpredictable
Capability Indexes are Predictable	Example 6.2	Example 6.4
Capability Indexes are Unpredictable	Example 6.3	Example 6.5

Figure 6.7: **The Relationship Between Charts for Process Data and Charts for Capability Ratios**

The second reason that capability indexes do not provide an effective way of monitoring your process is that the complex nature of the capability indexes makes them inherently more variable than the original data will be. This will make it harder for a signal of a change to show up in the capability indexes.

The third reason that capability indexes do not provide an effective way of monitoring your process is due to the structure of the *XmR* Chart. When capability indexes are placed on an *XmR* Chart the moving ranges will directly measure the variation of the index values. This variation is not the same as the variation in the process itself.

Thus, capability indexes are complex ratios which have their own inherent uncertainty and which only track the process in an indirect manner.

The purpose of placing the capability indexes on an *XmR* Chart is not to monitor the process, but rather to establish the amount of uncertainty that is inherent in the capability indexes themselves. Knowing the extent of this uncertainty will help you to avoid becoming excited about meaningless changes in the value of the capability index. It might even help you to calm down your customer when he thinks your numbers have taken a turn for the worse.

6.3 So What Do Capability Indexes Do?

- Capability indexes are descriptive measures which compare the space available with the space required for a predictable process.

- Capability Ratios do characterize the Elbow Room between specification limits for a predictable process.

- Centered Capability Ratios do characterize the process location relative to the specifications for a predictable process.

- Capability indexes that are greater than 1.00 do indicate that a predictable process will yield 100% conforming product as long as it continues to operate predictably.

- Capability indexes for a predictable process that are less than 1.00 have to be interpreted relative to specific knowledge of that process—they cannot be meaningfully used alone.

- Capability indexes do not characterize the fraction of nonconforming product for any process. Any conversion of a capability index into a fraction nonconforming will depend upon an assumed probability model. Change the model and you will change the fraction nonconforming.

- The best estimate of the fraction nonconforming is the ratio of the number of nonconforming items divided by the total number of items examined. Other estimates depend upon smoke and mirrors.

- Capability indexes do not tell you if a process is predictable or unpredictable.

- Capability Ratios do not characterize the elbow room for an unpredictable process.

- Centered Capability Ratios do not they characterize the location of an unpredictable process. Since both the location and dispersion of an unpredictable process are changing (unpredictably) any attempt to characterize these properties will be an exercise in futility.

- Finally, capability indexes do not tell you how to improve a process. They merely attempt to characterize how bad things are relative to meeting the Voice of the Customer. While there is a time and place for using such descriptive measures, you will not get anything done by concentrating all of your efforts on capability ratios.

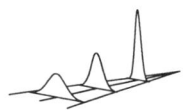

Chapter Seven

What Is Capability for Count Data?

What can you do if your data are count data and your customer is asking for capability indexes? Can you use the formulas in the previous chapters in some way? Just what do we mean when we talk about capability for count data?

To address these questions I will use an example based upon a letter I received.

Example 7.1: The Stamped Parts Data

Lee's process produced stamped parts, and the functionality of those parts depended more upon the overall shape of the part than it did on any one dimension of the part. While a part could have its several dimensions measured, it was hard to determine any relationship between these dimensions and the functionality of the part. In fact, some parts failed to work even when every dimension was within its tolerance range.

So Lee produced his parts and shipped them to the customer. Whenever the customer found a part that didn't work he would set it aside. When several parts had been set aside in this manner the customer would file a complaint and charge these defective parts back to the supplier. The report listed the number of failed parts, and the number of baskets of parts used. Since each basket contained 40 pieces, the numbers used were always a multiple of 40. Thus, the data provided by the customer complaints defined the quality of Lee's process. Lee's data consisted of counts of defective items and the counts of the number of items shipped. Since the customer sent in his reports at irregular intervals, the number of parts shipped would vary from report to report. A summary of the past 22 reports is shown below.

Report	No. Bad	No. Used	Percent Bad	Report	No. Bad	No. Used	Percent Bad
-1-	4	400	1.0	-12-	7	320	2.2
-2-	4	120	3.3	-13-	6	200	3.0
-3-	5	200	2.5	-14-	2	160	1.3
-4-	8	280	2.9	-15-	2	200	1.0
-5-	2	200	1.0	-16-	5	280	1.8
-6-	3	200	1.5	-17-	5	360	1.4
-7-	2	200	1.0	-18-	5	200	2.5
-8-	2	160	1.3	-19-	4	200	2.0
-9-	5	200	2.5	-20-	4	160	2.5
-10-	9	200	4.5	-21-	8	200	4.0
-11-	7	320	2.2	-22-	7	200	3.5

Lee placed these percentages on an *XmR* chart (see Figure 7.1). Interpreting this chart we see that Lee's process is averaging about 2% nonfunctional parts. In any one period of time this might vary from 0 to 5% nonfunctional, but the long-term average is close to 2% nonfunctional.

So should they be concerned about report number 10 which had 4.5% nonfunctional? Does 4.5% nonfunctional signify a change in Lee's process? Did things take a turn for the worse? The chart tells us that this value is within the bounds of routine variation, and should not be interpreted, by itself, as a signal of any change in the underlying process. Subsequent points on the chart bear out this interpretation.

The customer has set a standard that certifies the supplier as long as they do not have a failure rate in excess of 2.5%. This is the Voice of the Customer. The process behavior chart is the Voice of the Process. The heart of Lee's quandary is how to compute a capability ratio for this process.

Beyond Capability Confusion

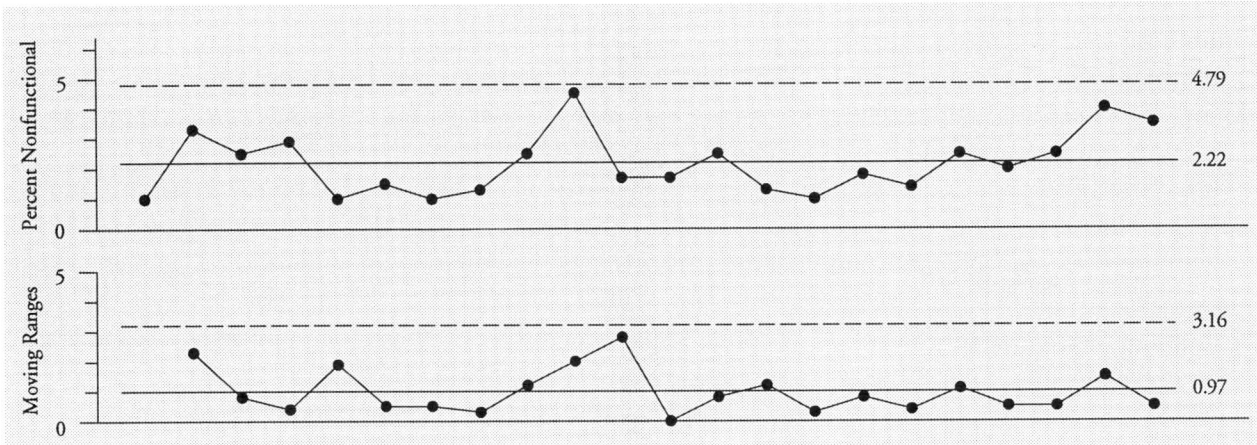

Figure 7.1: *XmR* Chart for the Percent Nonfunctional Stamped Parts

7.1 Can We Calculate a Centered Capability Ratio?

We seem to have everything we need to compute a Centered Capability Ratio: the process behavior chart provides us with estimates of the process average and dispersion, and the customer has given us an upper specification limit of 2.5 percent. So Lee could divide the average moving range of 0.97 by d_2 = 1.128 to obtain an estimate of his process dispersion: $Sigma(X)$ = 0.860. Then he could compute the difference between the upper specification limit and his process average: 2.5 − 2.22 = 0.28. Next he could divide this difference by 3 $Sigma(X)$ to get:

$$C_{pk} = \frac{0.28}{3\,(0.86)} = 0.11$$

Since we traditionally want the capability ratio to be greater than 1.00 this value of 0.11 would be considered to be a very bad number indeed.

However, this number is pure nonsense. It is a triumph of computation over common sense. It is meaningless, and the reasons why it is meaningless are outlined below.

7.2 Capability for Count Data

When working with measurements, specifications refer to individual characteristics of individual parts. The specifications apply to each part. In this situation, the question is "How likely is it that the parts will be within the specifications?" And this is the question that capability ratios were intended to address.

When working with counts the situation is somewhat different. In Lee's case, the customer specification is no more than 2.5% nonfunctional. Now what does this mean? Does this specification apply to each individual item? No. Each item is either functional or nonfunctional. No one item can be 2.5% nonfunctional! Thus we see the first major difference between specifications for measurement data and specifications for count data. *Specifications for count data must apply to the product stream, rather than to individual items in that stream.*

As soon as we recognize this difference between specifications for measurements and specifications for counts, we have to admit that the specification must be applied to a sequence of values. Now the next question is whether a specification for count data is intended to define an average, or a maximum, or a minimum.

If the specification is "no more than 2.5% nonfunctional on the average," then Lee's process is capable. Because Lee's process is predictable, we can say that, unless it changes for the worse, it will average about 2.2% nonfunctional, and will therefore satisfy this specification.

If the specification is "no more than 2.5% nonfunctional in any one period of time," then Lee's process is not capable of meeting this requirement. The process behavior chart tells us that Lee's process is not likely to produce more than 5% nonfunctional in any one period of time. So, if the 2.5% is an average, Lee is in good shape, but if it is a maximum, Lee is in trouble. Thus, before anything else happens, the specification needs to be clarified. Failure to clarify the nature of specifications for count data will inevitably result in arguments and beatings.

Next, because of the difference in the nature of specifications for count data, the traditional types of capability ratios do not make sense. They were focused on whether individual items were likely to meet the specifications. When you are counting items or counting events, each count represents something that actually happened. This difference in the nature of the data, along with the difference in the nature of the specifications, will make any type of capability ratio computation meaningless. So while we tortured the numbers above to come up with something that looked like a capability ratio, the result was, and will always be, meaningless.

With count data, the process behavior chart for the product stream defines the actual capability of a predictable process, or the potential capability of an unpredictable process. The central line defines what to expect over the long run, and the limits show the amount of short-term, routine variation about that long-term average that you should expect in practice.

Capability is a concept, rather than a computation. Learn the concept, make sense of your process, and use the computations when they are appropriate. Anything else is dangerous.

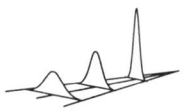

Chapter Eight

Why Specifications Don't Work

For the past two hundred years the traditional approach to quality has been summarized by the conformance to specifications. This approach underlies much of the work that we do, and it consumes much of our energy. And while the conformance to specifications is indeed important, there is more to lean manufacturing than merely meeting the specifications. As you will see in this and the next chapter, there is a new definition of quality that goes beyond meeting the specifications.

8.1 Why Specifications Don't Work

Consider the simple task of assembling a shaft and bearing. How do you make these two parts so that you end up with an assembly that works? Where do you set the target value for the shaft diameter? Where do you set the target value for the bearing diameter? And where do you set the specification limits for these two dimensions?

Say the target diameter for the shaft is 0.500 inches, and that the specifications are 0.495 to 0.505 inches. For the sake of simplicity, assume that all of our shafts are perfect cylinders—these shafts show no out-of-roundness, and they show no taper.

Because we have to allow some clearance for lubrication, say the target diameter for the bearings is 0.504 inches, with a tolerance of plus or minus 0.005 inches. Again, for the sake of simplicity, assume that we are making perfectly round bearings, with no taper and no out-of-roundness.

Since the process for making the shafts is different from that for making the bearings, we could represent the combined set of specifications as a rectangle on a two-dimensional plot. The square in Figure 8.1 defines those combinations of shaft diameters and bearing diameters that will be in conformance to the specifications.

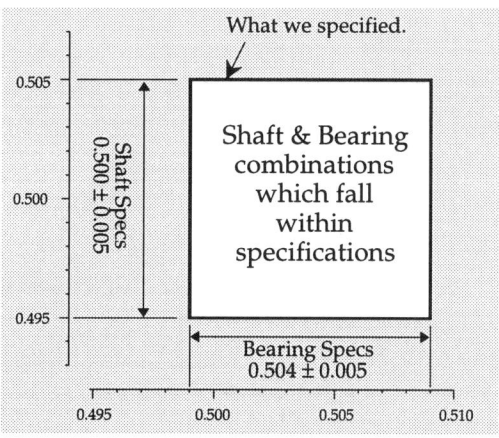

Figure 8.1: Specifications for Shafts and Bearings

The functionality of the assembly will depend upon the gap between the shaft and the bearing:

Gap = Bearing Diameter − Shaft Diameter

If the shaft diameter is greater than the bearing diameter, then the gap will be negative and the parts will not assemble. This will be called an interference fit.

If the gap is in the range of 0.001 to 0.003 inches the assembly will work as intended. Assemblies with gaps in this range will be called optimal fits.

Finally, if the gap exceeds 0.006 inches the assembly will leak lubricant. Assemblies with gaps greater than 0.006 inches will be called leakers. Thus, the job of producing assemblies that function is the job of fitting shafts to bearings such that the gap is in the range of 0.000 to 0.006 inches. This region intersects with the

Beyond Capability Confusion

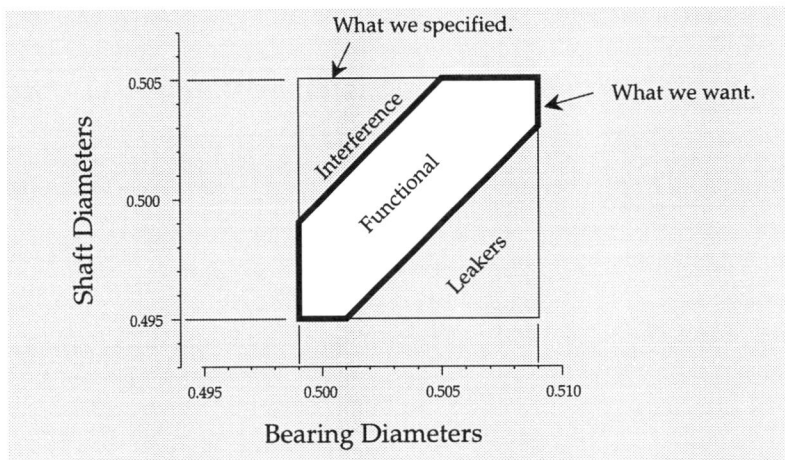

Figure 8.2: The Region of Functional Assemblies

specification rectangle to produce the diagonal band shown in Figure 8.2.

So the square is what we specified, and the diagonal band is what we want.

Is there any way to define the specifications for shaft diameters and bearing diameters to define a diagonal region in the plot of Figure 8.2? This is the problem of setting specifications. It is such an impossible problem that we usually give up at this point, pick some arbitrary specifications for shafts and bearings, cross our fingers and hope things will work out.

Now, in the interest of simplicity, we will assume that the production processes are predictable. The Bearing process is centered on a value of 0.504 inches, with a standard deviation of 0.002 inches. The Shaft process is centered on a value of 0.500 inches, with a standard deviation of 0.00133 inches. These two independent processes will result in a bivariate distribution with contours shown by the ellipse in Figure 8.3.

Thus, in Figure 8.3, the square is what we specified, the diagonal band is what we want, and the ellipse is what we are going to get. And therein lies the problem of manufacturing.

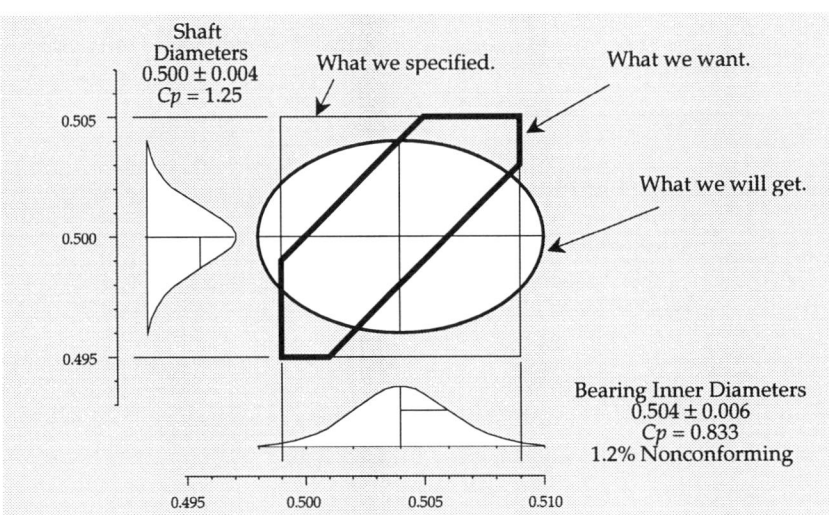

Figure 8.3: The Result of Two Predictable Processes

So what are you going to do? How are you going to operate your processes? How are you going to operate at assembly? Traditionally we look for quick fixes which seek to make the most of the current situation, or else we tweak the specifications to try and improve things slightly. And then we resign ourselves to trying to meet the production schedule with the inspect, sort, and rework operations.

8.2 Quick Fix Number One

With the predictable processes for making both the shafts and bearings shown in Figure 8.3, the gap in the initial assembly of the two parts may be characterized by the distribution shown in Figure 8.4. About 11% of the time a shaft will not fit into the bearing picked. When this happens the bearing will be put down and another bearing tried. About 1.2% of the time this second bearing will not fit and a third bearing will be tried. About one time in a thousand a fourth bearing will be fitted before the assembly will be completed. Thus, to get 1000 units to ship, the operator will have to perform, on the average, a total of 1246 operations as shown in Table 8.1.

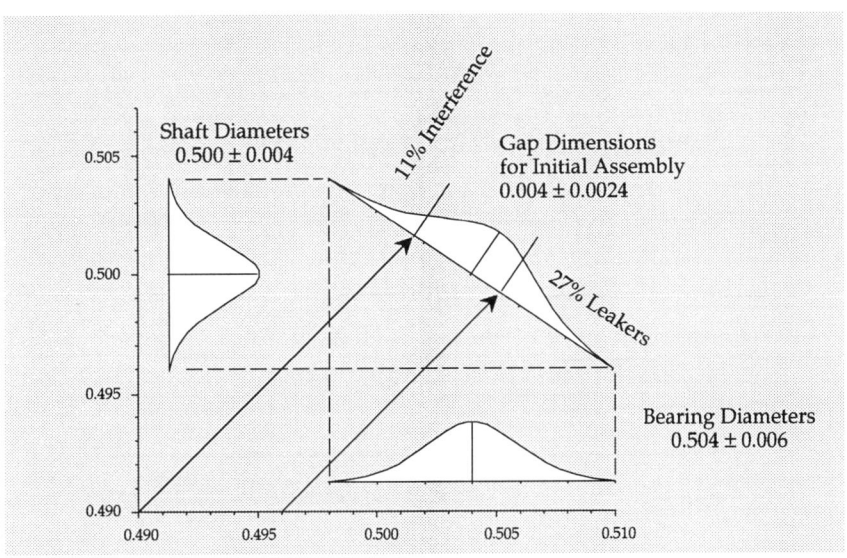

Figure 8.4: Gap Dimensions for Initial Assembly with Quick Fix No. 1

Table 8.1
Assemble 1000 units	1000 operations
Disassemble 110 interference fits	110 operations
Refit 110 units	110 operations
Disassemble 12 interference fits	12 operations
Refit 12 units	12 operations
Disassemble 1 interference fit	1 operation
Refit 1 unit	1 operation
Total:	1246 operations

However, with a 0.004 inch difference between the average shaft diameter and the average bearing diameter, about 370 of these 1000 units will be leakers, while only 220 will have an optimal fit.

Can we do better than this? Can we reduce the number of leakers? Well we could try to make smaller bearings.

Beyond Capability Confusion

8.3 Quick Fix Number Two

Say we shift the average Bearing Diameter down to 0.503 inches. With an average shaft diameter of 0.500 and an average bearing diameter of 0.503, the gaps may be characterized by the distribution shown in Figure 8.5. About 18% of the time a shaft will not fit into the bearing picked. When this happens the bearing will be put down and another bearing tried. About 3.2% of the time this second bearing will not fit and a third bearing will be tried. About six times in a thousand a fourth bearing will be fitted before the assembly will be completed, and about one time in a thousand a fifth fitting will be required. Thus, to get 1000 units to ship, the operator will have to perform, on the average, a total of 1438 operations as shown in Table 8.2.

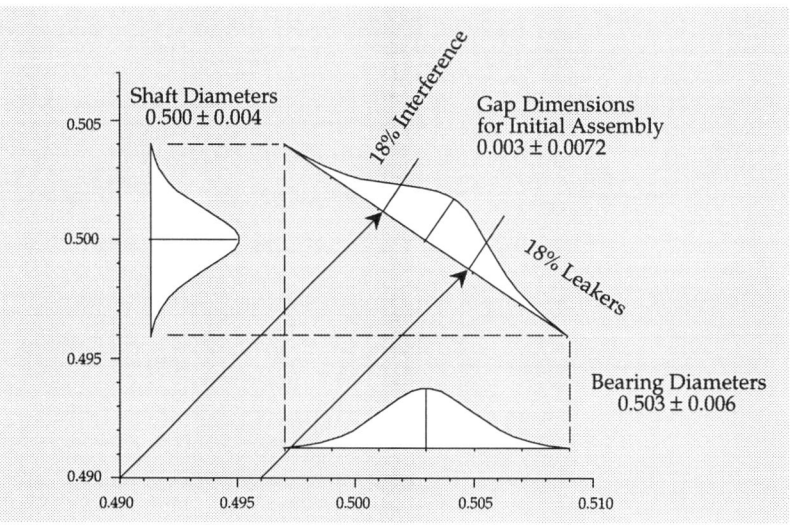

Figure 8.5: Gap Dimensions for Initial Assembly with Quick Fix No. 2

Table 8.2

Assemble 1000 units	1000 operations
Disassemble 180 interference fits	180 operations
Refit 180 units	180 operations
Disassemble 32 interference fits	32 operations
Refit 32 units	32 operations
Disassemble 6 interference fits	6 operations
Refit 6 units	6 operations
Disassemble 1 interference fit	1 operation
Refit 1 unit	1 operation
Total:	1438 operations

With a 0.003 inch difference between the average shaft diameter and the average bearing diameter, about 220 of these 1000 units will be leakers, while 280 will have an optimal fit. Both of these numbers are better than Quick Fix No. 1, but is it worth it? Quick Fix No. 1 required 1246 operations to get a total of 630 working units. This amounts to 1.98 operations per working unit. Essentially every working unit had to be assembled twice! Quick Fix No. 2 required 1438 operations to get a total of 780 working units, for a total of 1.84 operations per working unit. A slight improvement in productivity.

Can we do better than this? Well we could again reduce the average bearing diameter.

8.4 Quick Fix Number Three

Say we shift the average Bearing Diameter down to 0.502 inches. With an average shaft diameter of 0.500 and an average bearing diameter of 0.502, the gaps may be characterized by the distribution shown in Figure 8.6. About 27% of the time a shaft will not fit into the bearing picked. When this happens the bearing will be put down and another bearing tried. About 7.3% of the time this second bearing will not fit and a third bearing will be tried. About 2.0% of the time a fourth bearing will be fitted before the assembly will be completed. A fifth fitting will be needed about 0.5% of the time, and a sixth fitting will be needed about 0.1% of the time. Thus, to get 1000 units to ship, the operator will have to perform, on the average, a total of 1738 operations as shown in Table 8.3.

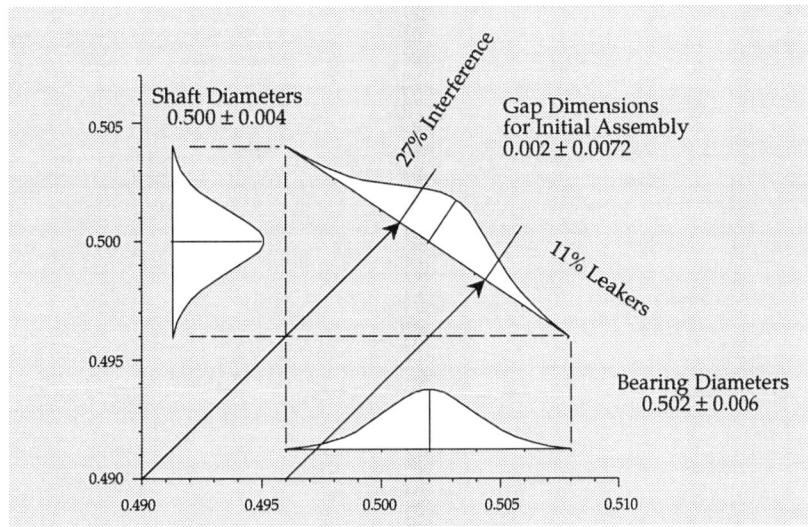

Figure 8.6: Gap Dimensions for Initial Assembly with Quick Fix No. 3

Table 8.3

Assemble 1000 units	1000 operations
Disassemble 270 interference fits	270 operations
Refit 270 units	270 operations
Disassemble 73 interference fits	73 operations
Refit 73 units	73 operations
Disassemble 20 interference fits	20 operations
Refit 20 units	20 operations
Disassemble 5 interference fits	5 operations
Refit 5 units	5 operations
Disassemble 1 interference fit	1 operation
Refit 1 unit	1 operation
Total:	1738 operations

With a 0.002 inch difference between the average shaft diameter and the average bearing diameter, about 150 of these 1000 units will be leakers, while 320 will have an optimal fit. This is better than Quick Fixes One and Two. However, we have performed 1738 operations to get 850 working units, for a total of 2.04 operations per working unit. This value is slightly higher than the preceding approaches. We are still essentially assembling each working unit twice.

What would happen if we further reduced the difference between the shaft and bearing diameters?

8.5 Quick Fix Number Four

Say we shift the average Bearing Diameter down to 0.501 inches. With an average shaft diameter of 0.500 and an average bearing diameter of 0.501, the gaps may be characterized by the distribution shown in Figure 8.7. About 39% of the time a shaft will not fit into the bearing picked. When this happens the bearing will be put down and another bearing tried. About 15.2% of the time this second bearing will not fit and a third bearing will be tried. About 5.9% of the time a fourth bearing will be fitted before the assembly will be completed. A fifth fitting will be needed about 2.3% of the time, a sixth fitting will be needed about 0.9% of the time, a seventh fitting will be needed about 0.3% of the time, and an eighth fitting will be needed about 0.1% of the time. Thus, to get 1000 units to ship, the operator will have to perform, on the average, a total of 2274 operations as shown in Table 8.4.

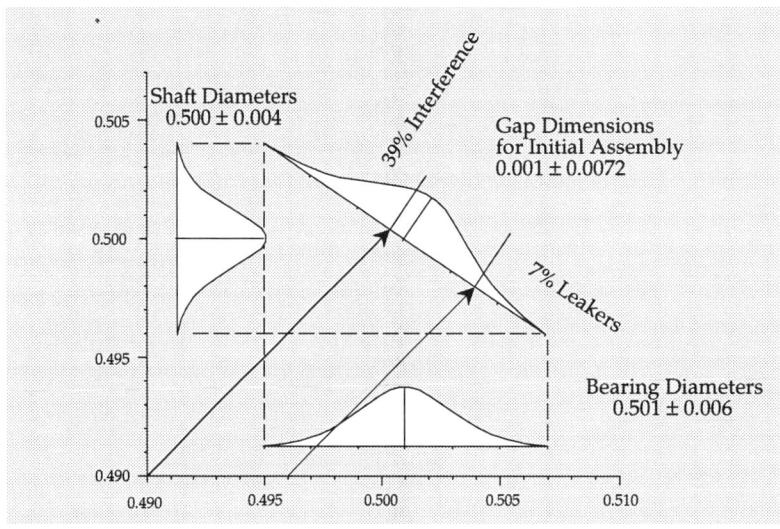

Figure 8.7: Gap Dimensions for Initial Assembly with Quick Fix No. 4

With a 0.001 inch difference between the average shaft diameter and the average bearing diameter, about 110 of these 1000 units will be leakers, while 370 will have an optimal fit. With this approach we have performed 2274 operations to get 890 working units, for a total of 2.56 operations per working unit. While the number of units with optimal fit is up, and the leakers are down, the labor costs per unit are up as well. However, our customer is pressuring us to quit shipping those leakers, so we continue to tinker.

Table 8.4

Assemble 1000 units	1000 operations
Disassemble 390 interference fits	390 operations
Refit 390 units	390 operations
Disassemble 152 interference fits	152 operations
Refit 152 units	152 operations
Disassemble 59 interference fits	59 operations
Refit 59 units	59 operations
Disassemble 23 interference fits	23 operations
Refit 23 units	23 operations
Disassemble 9 interference fits	9 operations
Refit 9 units	9 operations
Disassemble 3 interference fits	3 operations
Refit 3 units	3 operations
Disassemble 1 interference fit	1 operation
Refit 1 unit	1 operation
Total:	2274 operations

8.6 Quick Fix Number Five

Say we shift the average Bearing Diameter down to 0.500 inches. With an average shaft diameter of 0.500 and an average bearing diameter of 0.500, the gaps may be characterized by the distribution shown in Figure 8.8. About 50% of the time a shaft will not fit into the bearing picked. When this happens the bearing will be put down and another bearing tried. Following the scenario shown in Table 8.5, the operator will have to perform, on the average, a total of 3002 operations to get 1000 units to ship.

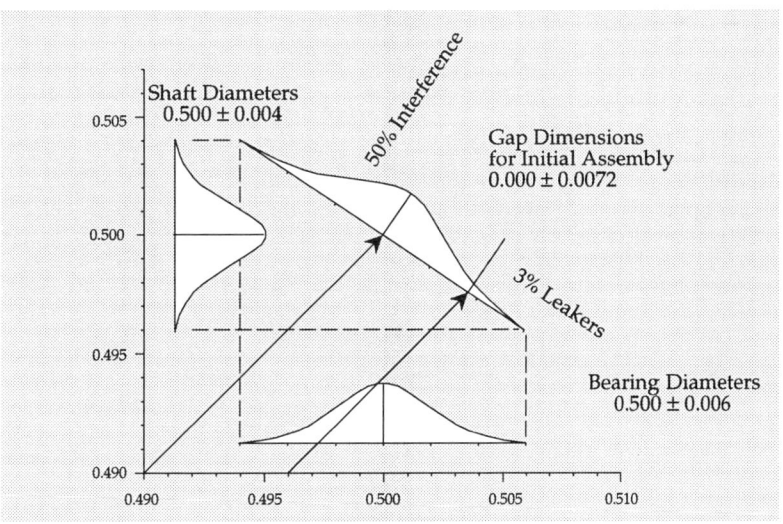

Figure 8.8: Gap Dimensions for Initial Assembly with Quick Fix No. 5

Table 8.5

Assemble 1000 units	1000 operations
Disassemble 500 interference fits	500 operations
Refit 500 units	500 operations
Disassemble 250 interference fits	250 operations
Refit 250 units	250 operations
Disassemble 125 interference fits	125 operations
Refit 125 units	125 operations
Disassemble 63 interference fits	63 operations
Refit 63 units	63 operations
Disassemble 32 interference fits	32 operations
Refit 32 units	32 operations
Disassemble 16 interference fits	16 operations
Refit 16 units	16 operations
Disassemble 8 interference fits	8 operations
Refit 8 units	8 operations
Disassemble 4 interference fits	4 operations
Refit 4 units	4 operations
Disassemble 2 interference fits	2 operations
Refit 2 units	2 operations
Disassemble 1 interference fit	1 operation
Refit 1 unit	1 operation
Total:	3002 operations

Of these 1000 units about 60 will be leakers, while 380 will have an optimal fit. This was accomplished with 3.19 operations per working unit—in effect, each working unit has now been assembled three times!

Clearly we have reached the point of diminishing returns here. Is there some other approach we could use? What about sorting to fit?

8.7 Quick Fix Number Six

Here we consider the favorite solution of mass production—pre-sorting the parts so that they will fit when assembled. Let us shift the Bearing Diameter process back to an average of 0.502 inches to reduce the interference fit problem, and instead of the random assembly of shafts and bearings we pre-sort the bearings and shafts into categories and assemble parts from corresponding categories. We could sort the bearings into three ranges: low (0.497 to 0.501), medium (0.501 to 0.504), and high (0.504 to 0.507). With a properly constructed plug gauge this sorting could be done with a single operation per bearing. The shafts could be sorted into low (0.496 to 0.498), medium (0.498 to 0.501), and high (0.501 to 0.504). With the predictable processes assumed for the shafts and bearings this sorting would require about 1.93 operations per piece. (If a shaft fits a Low Range Gauge, it is placed in the low category. If it does not fit the Low Range Gauge, it will have to be tried with the Mid Range Gauge. About 93% of the shafts will require this second operation.) Thus we are up to 2.93 operations per assembly *before we even begin to assemble the units*. However, with these ranges we will have virtually no interference fits, and virtually no leakers (see Figure 8.9).

Table 8.6

Sort 1000 shafts into low or other category	1000 operations
Sort 930 shafts into medium or high category	930 operations
Sort 1000 bearings into Low, med. high cat.	1000 operations
Assemble 1000 units	1000 operations
Total:	3930 operations

The operator will have to perform 3930 operations to get 1000 units to ship. None of these units will be leakers, and about 460 of them will have optimal fit. While this should make our customer happy, it has cost us a total of 3.93 operations per unit. We shipped no unit until it was effectively assembled four times!

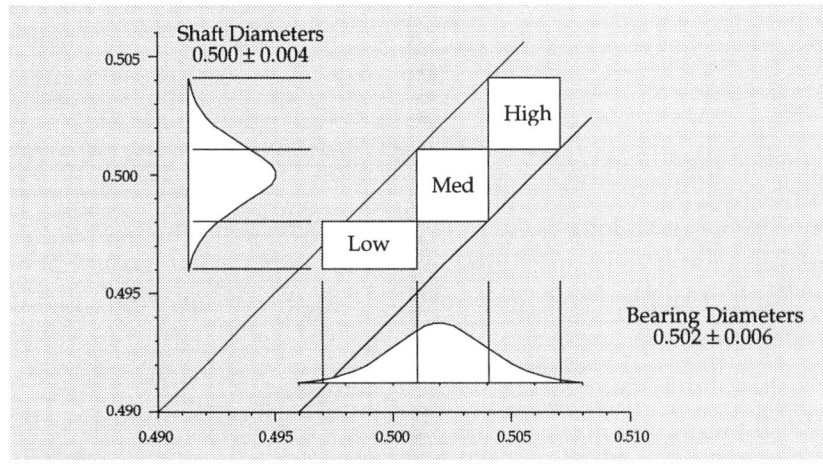

Figure 8.9: Gap Dimensions for Assembly with Quick Fix No. 6

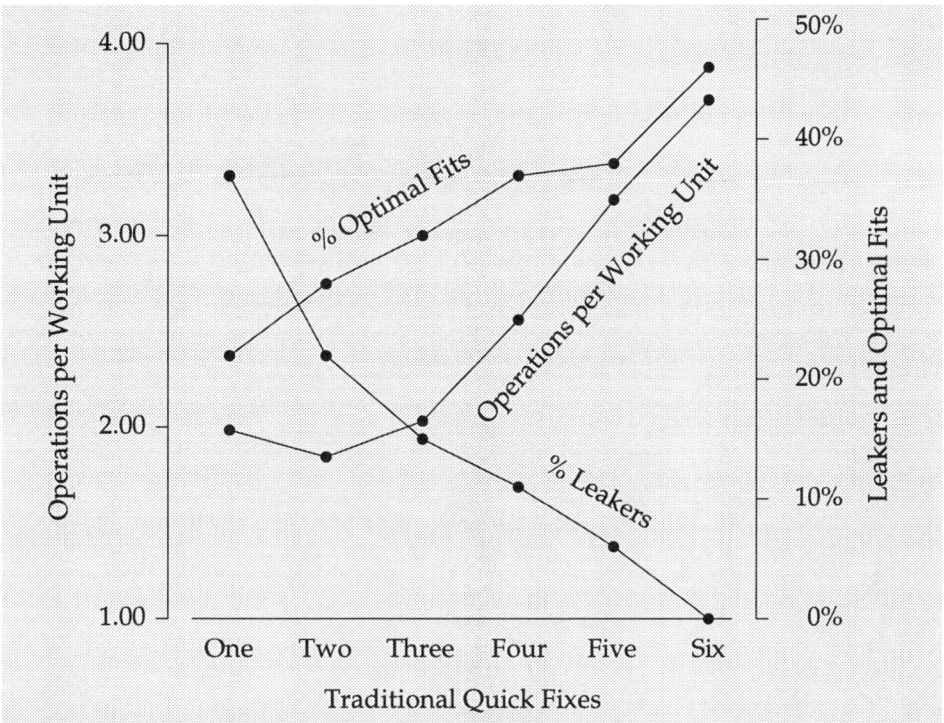

Figure 8.10: The Traditional View of the Relationship Between Quality and Productivity

A summary of these traditional quick fixes is shown in Figure 8.10. There the final percentage of leakers, the final percentage of optimal fits, and the number of operations per working unit are plotted together. As we move from Quick Fix One to Quick Fix Six the percentage of leakers drops, and the percentage of units with optimal fit climbs (increasing quality). However, these improvements were only accomplished by dramatically increasing the labor content of each unit (decreasing productivity).

Given these approaches, is it any wonder that the traditional view is that quality and productivity are contrary to each other? For over 200 years the Western Approach to Quality has been: "burn the toast and scrape it."

But there is a better way—stop burning the toast!

8.8 The New Approach to Production: Continual Improvement

The Conformance to Specifications Concept of Quality is focused on achieving a minimum level of performance. Continual Improvement is focused on getting what your process can deliver. We are not going to ask it to do more than it is capable of doing, but neither are we going to be content to let it do less than it is capable of doing.

The methodology of getting any process to perform up to its potential is to use process behavior chart to track the process behavior over time. In the natural course of events, all processes will be subject to changes. By using process behavior charts to detect these unknown changes, and by investigating the causes of these changes, you can discover how to improve your process with little or no capital expenditures. Typically, as you find and remove assignable causes of excessive variation, your process variation will drop to one-half, one-third, or even one-fourth of its previous value. This reduced variation then simplifies all subsequent operations, resulting in additional savings, improved quality, and improved

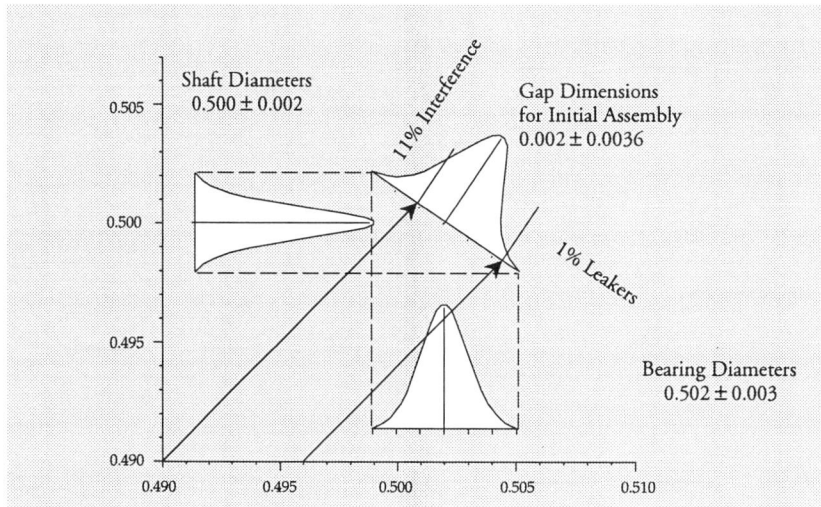

Figure 8.11: Gap Dimensions for Initial Assembly with Continual Improvement

productivity.

To see just what this new approach can do, consider the effect of reducing the variation in both the shafts and the bearings by 50%. Assume the Shaft process is predictable, with an average shaft diameter of 0.500 inches, and a standard deviation of 0.000667 inches. Also assume that the Bearing process is predictable, with an average Bearing Diameter of 0.502 inches, and a standard deviation of 0.001 inches. With these values, random assembly of shafts and bearings will result in about 11% interference fits, and a second fitting will be required. About 1.2% of the time a third fitting will be needed. About one time in a thousand a fourth bearing will be fitted before the assembly will be completed. Thus, to get 1000 units to ship, the operator will have to perform, on the average, a total of 1246 operations as shown in Table 8.7.

Table 8.7

Assemble 1000 units	1000 operations
Disassemble 110 interference fits	110 operations
Refit 110 units	110 operations
Disassemble 12 interference fits	12 operations
Refit 12 units	12 operations
Disassemble 1 interference fit	1 operation
Refit 1 unit	1 operation
Total:	1246 operations

However, unlike Quick Fix No. One which also required 1246 operations per 1000 units on the average, we will now have only 10 leakers out of the 1000 units. In addition, 510 of the 1000 units will have optimal fit. Thus, 1246 operations to get 990 working units yields a total of 1.26 operations per unit. We get the *best quality* with the *least effort* by learning how to *get the most out of our process*.

The new approach is compared with the traditional quick fixes in Figure 8.12. Continual Improvement results in more optimal fits, essentially no leakers, and the lowest labor content of any of the approaches considered.

Of course many production problems are more complex than the simple situation used here. But if the specification approach does not work in a simple case, is it likely to do any better in a complex case?

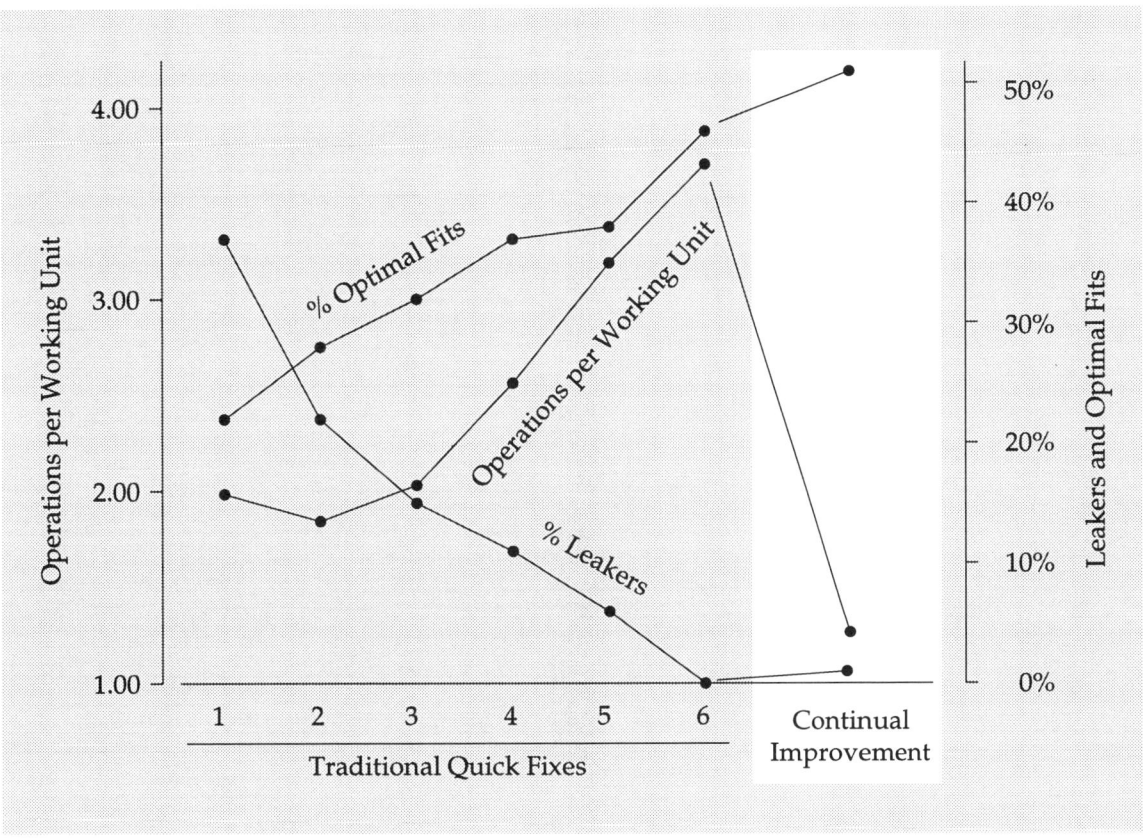

Figure 8.12: The Difference Between the Quick Fixes and Continual Improvement

The Conformance to Specifications Concept of Quality is obsolete. Specifications may be used to scrape the toast, but they do not create quality. All specification-based definitions of quality completely miss the point. Whenever you focus on specifications you will always end up arguing about how good the parts have to be. When you focus on getting the most out of your processes, you will get more uniform product, with less effort and greater productivity. World-class quality is no longer a matter of conformance to specifications. As we shall see in the next chapter, world-class quality is now defined to be on-target with minimum variance. Those who do not learn the difference between the specification-based approaches to quality and the continual improvement approach will only become increasingly noncompetitive.

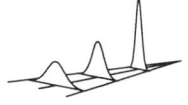

Chapter Nine

The Average Cost-of-Use

The problem with focusing solely on the specifications was described in the previous chapter. In this chapter an alternative way of thinking about, and measuring, your processes will be presented.

9.1 Variation Always Creates Costs

"While we were busy arguing about how good the parts had to be,
they were working hard at making all the parts exactly the same." *

Why would John Betti, then Vice-President of Powertrain Operations at Ford Motor Company, be concerned about their competition working on making all the parts exactly the same? Wouldn't this just make things more expensive? After all, doesn't it cost more to take variation out of the product? "We can't worry about the variation of the product, we have a production schedule to meet—so just speed up the line, make more parts, and then we can sort out the conforming product later. And anyway, we can always rework the reject parts and sell them as replacement parts."

This attitude is typical of a mass production environment. Variation is viewed as inherent in the process, and so we have to take steps to deal with variation after the fact. And these actions always have a price tag, which is why quality and productivity are commonly viewed as opposites.

But, as was outlined in the previous chapter, there is another way of dealing with variation. Instead of paying someone to create the variation and then paying someone else to remove that variation, you can learn how to remove the variation up front. By reducing variation at the source you can reduce subsequent costs while increasing product quality. The further upstream you work to reduce variation, the lower the costs of variation will be. And that is what this chapter is about—the costs of variation.

Variation always creates costs. While we have known this for a long time, we have been unable to do anything about it because we have not had effective ways of working with variation. And since we have been fully occupied in trying to meet the shipping schedule in spite of the bottlenecks of inspection and rework, we have tended to forget this basic truth that variation always creates costs.

Moreover, we have always known that actions taken to deal with variation after the fact will increase the costs. However, at first the economies of mass production were sufficient to absorb these extra costs. Later they simply became a way of life, so no one questioned if there was a better way. Now that a better way is known to exist, it is time to go beyond the traditional thinking that grew up with mass production.

It should be self-evident that actions that reduce variation at the source will reduce subsequent costs while increasing product quality. However, without a methodology for dealing with variation at the source, this was a strategy that remained out of reach. Today there is an easy, effective, and proven methodology for reducing variation at the source—the use of process behavior charts for continual improvement. While an explanation of this methodology is beyond the scope of this book, this chapter will provide a metric for the journey of continual improvement.

---

* John Betti said this about transaxles produced by Mazda. For the rest of the story see Chapter 11 of *The Deming Dimension* by Henry Neave.

Beyond Capability Confusion

9.2 The Costs of Using Conforming Product

Yes, there are costs associated with using conforming product.

In the past we have tended to focus on the costs associated with nonconforming product. We have totaled up the failure costs, the appraisal costs, and the prevention costs, and lumped them all together into something that we called the cost of quality. But the cost of quality does not incorporate all of the costs associated with using conforming product.

To understand these costs recall the example in the previous chapter. There we assumed that both the shaft process and the bearing process were essentially capable. In spite of this there was a huge labor burden necessary to handle the problem of interference fits. In some cases this burden was two to three times the nominal labor burden needed to assemble the product. The components were conforming, but the extra work, and the extra inventory required to get them to assemble, were all costs of using conforming product. The various quick fixes could not overcome this problem because they all attempted to deal with variation after the fact.

In addition to this refitting of the parts when an interference fit occurs, there is the extra expense of leak testing all of the units produced. While this was not specifically counted in the previous chapter, it is still a cost associated with using these conforming shafts and bearings.

And then there is the cost of repairing the leakers. This will require another work area, more workers, special equipment, and additional inventory. In addition, the Environmental Protection Agency has made it more costly to clean up the mess associated with this repair operation. All of these costs are costs of using the conforming shafts and bearings. While the quick fixes tended to reduce the number of leakers, they did so by shifting the labor back upstream. When assembly and rework are separate operations, with separate managers, you will have a real conflict as each tries to look better by shifting work to the other department (an indirect cost of using the conforming shafts and bearings).

Finally, after the assembly and refitting, after the leak-testing and repair, the units get used. Those units with an optimal gap will probably operate as intended. Those with too tight a gap, or with too loose a gap, will tend to be the early failures. These may be internal failures, or failures under warranty, but in the end, they add to the cost of using the conforming shafts and bearings.

In the shaft and bearing example there was no sorting of the bearings and shafts to separate the conforming from the nonconforming—they were essentially all conforming. However, such screening would also be a cost associated with using the conforming product. While some of these costs are captured in the traditional cost of quality accounting, most are not.

In contrast with the Quick Fixes in the previous chapter, consider what would happen if we could produce all of the bearings at the target, and all of the shafts at the target. If we could make 100 percent of the shafts at 500 mils, and 100 percent of the bearings at 502 mils, then we would get 100 percent optimal fits on initial assembly. There would be no leakers, and no interference fits. We would need 1000 operations to get 1000 units to ship, and since every unit would have an optimal fit there would be no need to perform 100% leak-testing, nor any repair line. This optimal assembly scenario is the baseline. Every thing that raises the cost above this optimal level is a cost of using conforming product—a cost of variation.

All of these costs are costs of using conforming product. All of these costs are real. They all affect the bottom line, even though we bury them in different places and in different ways. This is why John Betti was concerned about the competition making all the parts exactly the same, because it resulted in a better product at a lower cost. In that example, the warranty costs for the competitor's units were one-tenth of the warranty costs for the same units from Ford, and the customers who had unknowingly bought cars with the Mazda transaxles were detectably happier with their Escorts than the others, which is an intangible cost of using conforming product.

While you might have trouble evaluating some of these costs associated with using conforming product, it will generally be sufficient to use those that you can identify. They will usually be enough to justify the effort required to reduce variation at the source.

9.3 The Average Cost-of-Use

The costs of using conforming product can be characterized by a simple expression. This quantity is the Average Cost-of-Use. It is the average cost of using conforming product per unit of production, expressed in dollars. It could be computed for each product characteristic for a given production process and can be used to understand the costs of variation associated with each of those characteristics.

To compute the Average Cost-of-Use for a given product characteristic you will need to know certain things about the product and the process that produces it. In particular, you will need to know:

1. The target value for the product characteristic. This value will often be the mid-point of the specification limits.
2. The cost of scrapping or reworking an item, C_{scrap}.
3. That point at which you would either scrap or rework the item, X_{scrap}.
4. The average value for the product stream. This would typically be the Grand Average from a process behavior chart.
5. A estimate of the variation in the product stream. Again, this would typically come from a process behavior chart:

$$Sigma(X) = \frac{\bar{R}}{d_2}$$

Given these five values the simplest version of the Average Cost-of-Use may be computed as:

$$ACU = K \left\{ [Sigma(X)]^2 + [\bar{\bar{X}} - target]^2 \right\}$$

where K is a constant that converts the quantity in brackets into monetary units. A value for this constant may be found using the formula:

$$K = \frac{C_{scrap}}{[X_{scrap} - target]^2}$$

This constant K will always have the form of (monetary units) divided by (measurement units squared). The expression in brackets will always have the form of (measurement units squared). Thus, the product of these two values will result in a number expressed in monetary units. This will be the Average Cost-of-Use per unit of production. This cost is the average cost per unit that can be attributed to the fact that all of the units are not exactly at the target value. It is the cost of variation about the target. It is the cost of using conforming product.

An explanation of the origin of these equations will be given in the following sections of this chapter, but first we will look at four examples of the use of these formulas.

Example 9.1: The Yield Data

In Example 5.2 we found the Yield process to be on the Brink of Chaos—unpredictable but conforming. The *XmR* chart for this process is shown in Figure 9.1. The six spikes in the moving range chart identify six shifts in the process level. These shifts occurred between the different production runs. The *X* chart in Figure 9.1 shows these seven production runs with separate limits.

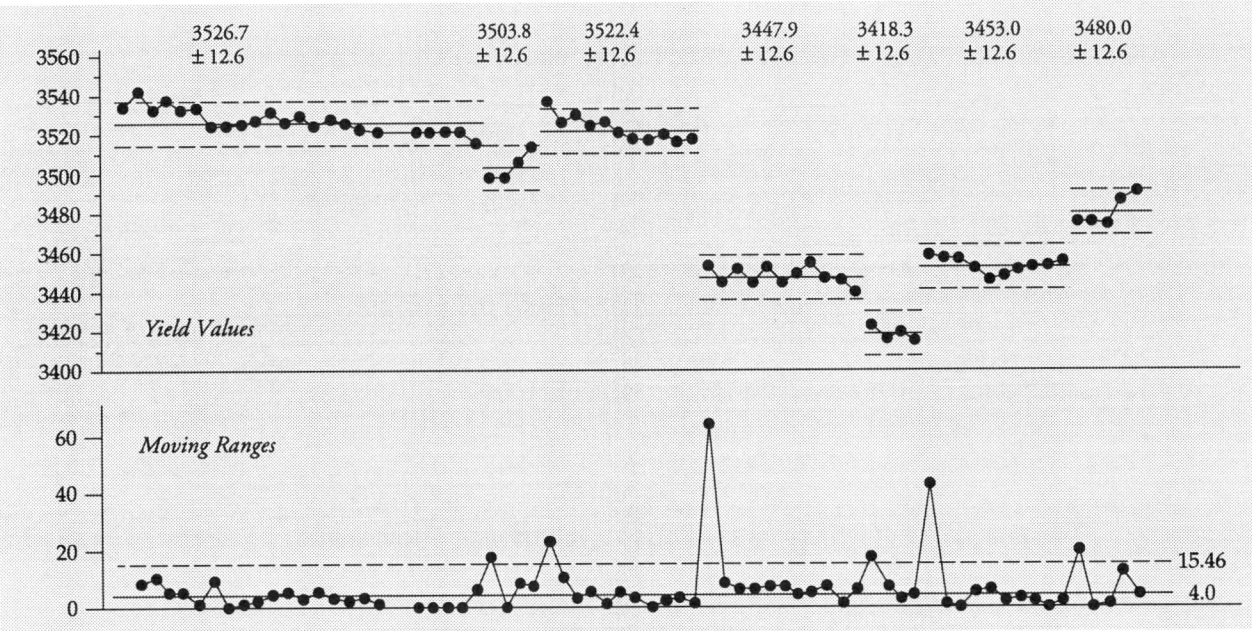

Figure 9.1: The *XmR* Chart for the Yield Data

The specification limits are 3410 to 3550. Let us assume that the target value for yield is the midpoint of the specifications, which is 3480 units. Also, let us assume that the cost of scrapping a single nonconforming item is $1.00. The scrap points would be the specification limits. Using these values we find a value for *K* to be:

$$K = \frac{C_{scrap}}{[X_{scrap} - target]^2} = \frac{\$\,1.00}{(3550\text{ units} - 3480\text{ units})^2} = \frac{\$\,0.000204}{\text{unit}^2}$$

The process behavior chart reveals that the big changes occur between the short production runs. Therefore we shall compute the Average Cost-of-Use for each of these seven production runs. Dividing the median moving range by 0.954 yields a value of *Sigma(X)* of 4.2 units. We shall use this value for all seven production runs because the moving range chart suggests that this would make sense in this case.

For Production Run One the average is 3526.7. Thus, for this run the Average Cost-of-Use is:

$$ACU = K \left\{ [Sigma(X)]^2 + [\bar{\bar{X}} - target]^2 \right\}$$

$$= \frac{\$\,0.000204}{\text{unit}^2} \left\{ [4.2\text{ units}]^2 + [3526.7\text{ units} - 3480\text{ units}]^2 \right\}$$

$$= \frac{\$\,0.000204}{\text{unit}^2} (2198.5\text{ units squared}) = \$\,0.4485 \text{ per item}$$

For Production Run Two the average is 3503.8. Thus, for this run the Average Cost-of-Use is:

$$ACU = K \left\{ [Sigma(X)]^2 + [\bar{\bar{X}} - target]^2 \right\}$$

$$= \frac{\$\,0.000204}{unit^2} \left\{ [4.2 \text{ units}]^2 + [3503.8 \text{ units} - 3480 \text{ units}]^2 \right\}$$

$$= \frac{\$\,0.000204}{unit^2} (584.1 \text{ units squared}) = \$\,0.1192 \text{ per item}$$

For Production Run Three the average is 3522.4. Thus, the Average Cost-of-Use is:

$$ACU = K \left\{ [Sigma(X)]^2 + [\bar{\bar{X}} - target]^2 \right\}$$

$$= \frac{\$\,0.000204}{unit^2} \left\{ [4.2 \text{ units}]^2 + [3522.4 \text{ units} - 3480 \text{ units}]^2 \right\}$$

$$= \frac{\$\,0.000204}{unit^2} (1815.4 \text{ units squared}) = \$\,0.3703 \text{ per item}$$

For Production Run Four the average is 3447.9. Thus, the Average Cost-of-Use is:

$$ACU = K \left\{ [Sigma(X)]^2 + [\bar{\bar{X}} - target]^2 \right\}$$

$$= \frac{\$\,0.000204}{unit^2} \left\{ [4.2 \text{ units}]^2 + [3447.9 \text{ units} - 3480 \text{ units}]^2 \right\}$$

$$= \frac{\$\,0.000204}{unit^2} (1048.1 \text{ units squared}) = \$\,0.2138 \text{ per item}$$

For Production Run Five the average is 3418.3. Thus, the Average Cost-of-Use is:

$$ACU = K \left\{ [Sigma(X)]^2 + [\bar{\bar{X}} - target]^2 \right\}$$

$$= \frac{\$\,0.000204}{unit^2} \left\{ [4.2 \text{ units}]^2 + [3418.3 \text{ units} - 3480 \text{ units}]^2 \right\}$$

$$= \frac{\$\,0.000204}{unit^2} (3824.5 \text{ units squared}) = \$\,0.7802 \text{ per item}$$

For Production Run Six the average is 3453.0. Thus, the Average Cost-of-Use is:

$$ACU = K \left\{ [Sigma(X)]^2 + [\bar{\bar{X}} - target]^2 \right\}$$

$$= \frac{\$\,0.000204}{unit^2} \left\{ [4.2 \text{ units}]^2 + [3453.0 \text{ units} - 3480 \text{ units}]^2 \right\}$$

$$= \frac{\$\,0.000204}{unit^2} (746.6 \text{ units squared}) = \$\,0.1523 \text{ per item}$$

And for Production Run Seven the average is 3480.0. Thus, the Average Cost-of-Use is:

$$ACU = K \left\{ [Sigma(X)]^2 + [\bar{\bar{X}} - target]^2 \right\}$$

$$= \frac{\$\ 0.000204}{unit^2} \left\{ [4.2\ units]^2 + [3480.0\ units - 3480\ units]^2 \right\}$$

$$= \frac{\$\ 0.000204}{unit^2} (17.6\ units\ squared) = \$\ 0.0036\ per\ item$$

Since each point on the *XmR* Chart in Figure 9.1 represents approximately 400 items produced, we can extend these Average Costs-of-Use to characterize the costs of letting this process drift around from production run to production run.

Table 9.1 Costs of Using Conforming Product

Production Run	Approximate No. of Units Made	Average Cost-of-Use	Costs of Using Conforming Product
1	9200	$ 0.4485	$ 4126.20
2	1600	$ 0.1192	$ 190.72
3	4400	$ 0.3703	$ 1629.32
4	4400	$ 0.2138	$ 940.72
5	1600	$ 0.7802	$ 1248.32
6	4000	$ 0.1523	$ 609.20
7	2000	$ 0.0036	$ 7.20

The last production run shows what this process is capable of doing. The first six runs resulted in approxiamtely $8600.00 of excess costs because the manufacturer was not getting what his process was capable of producing. As far as we can tell from these data none, or very few, of the 27,200 units produced were nonconforming. Yet, the wide specification limits, and the poor job of setting up each production run, results in variation that creates substantially higher costs of use than are necessary. Earlier we assumed that each item cost the producer about $1.00. This means that approximately 30 percent of the cost of these items is tied up in the cost of variation. What is going to happen to this producer when he has to compete with someone that practices lean production techniques?

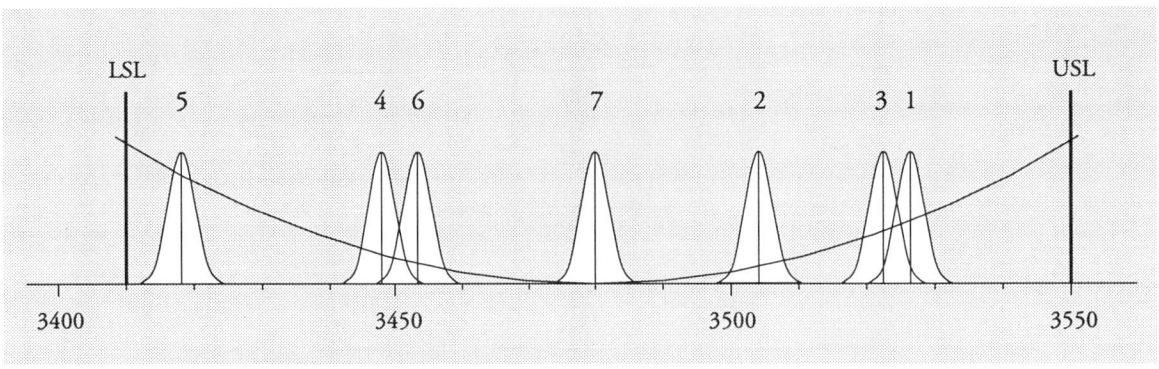

Figure 9.2: The Seven Production Runs for the Yield Data

Example 9.2: The Rheostat Knob Data

In Example 4.1 we found the Rheostat Knob process to be predictable and capable. The midpoint of the specifications is 0.140 inches, which we will assume to be the target value. Say the cost of scrapping a knob is $0.05, and that the scrap point would be either specification limit of 0.125 inches or 0.155 inches. Thus a value for K is found to be:

$$K = \frac{C_{scrap}}{[\,X_{scrap} - \text{target}\,]^2} = \frac{\$0.05}{(0.155 \text{ inches} - 0.140 \text{ inches})^2} = \frac{\$\,2.222}{\text{inch}^2}$$

The process behavior chart shows a Grand Average of 0.14063 inches, and an Average Range of 0.00856 inches. With subgroups of size $n = 5$, the value for d_2 is 2.326. Thus $Sigma(X)$ is 0.00368 in. Combining these with the value for K we get the Average Cost-of-Use for this process to be:

$$ACU = K\,\{\,[\,Sigma(X)\,]^2 + [\,\bar{\bar{X}} - \text{target}\,]^2\,\}$$

$$= \frac{\$\,2.222}{\text{inch}^2}\,\{\,[\,0.00368 \text{ in.}\,]^2 + [\,0.14063 \text{ in.} - 0.140 \text{ in.}\,]^2\,\}$$

$$= \frac{\$\,2.222}{\text{inch}^2}\,(0.00001394 \text{ inches squared}) = \$\,0.000031$$

This is the Average Cost-of-Use per knob. This very small value suggests that there is little additional benefit to be had by working to improve this process. It is already operating on-target, well inside the specifications, and up to its potential.

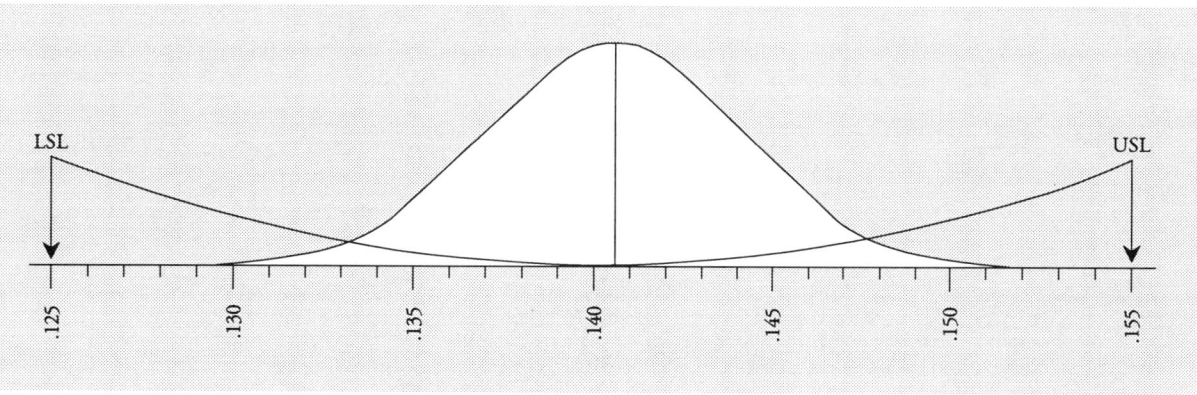

Figure 9.3: The Rheostat Knob Process

Example 9.3: The Ball-Joint Socket Thickness Data

In Example 4.2 we found the Ball-Joint Thickness process to be predictable and capable. The midpoint of the specifications is 7.5 units, which we will assume to be the target value. Say the cost of scrapping a socket is $0.25, and that the scrap point would be at either of the specification limits of 0 units or 15 units. Thus a value for K is found to be:

$$ K = \frac{C_{scrap}}{[X_{scrap} - target]^2} = \frac{\$\,0.25}{(15 \text{ units} - 7.5 \text{ units})^2} = \frac{\$\,0.00444}{\text{unit}^2} $$

The process behavior chart in Figure 4.10 shows a Grand Average of 4.66 units, and an Average Range of 3.71 units. With subgroups of size $n = 4$, the value for d_2 is 2.059, giving a $Sigma(X)$ value of 1.80 units. Combining these with the value for K we get the following Average Cost-of-Use for this process:

$$ \begin{aligned} ACU &= K \left\{ [Sigma(X)]^2 + [\bar{\bar{X}} - target]^2 \right\} \\ &= \frac{\$\,0.00444}{\text{unit}^2} \left\{ [1.80 \text{ units}]^2 + [4.66 \text{ units} - 7.5 \text{ units}]^2 \right\} \\ &= \frac{\$\,0.00444}{\text{unit}^2} (11.31 \text{ units squared}) = \$\,0.0502 \text{ per piece} \end{aligned} $$

This Average Cost-of-Use is over 20 percent of the cost of scrap. So even though none of the items are nonconforming, there is still a substantial cost associated with using the conforming items. By getting this dimension on target they could reduce the Average Cost-of-Use for this characteristic by the following amount:

$$ \frac{\$\,0.00444}{\text{unit}^2} \left\{ [4.66 \text{ units} - 7.5 \text{ units}]^2 \right\} = \$\,0.0358 \text{ per piece} $$

Thus, over 70 percent of the original cost of use is due to the process being off-target. Correcting this off-target condition will save over $4000.00 in six months time. This benefit can be set against the cost of correcting the off-target condition of this process.

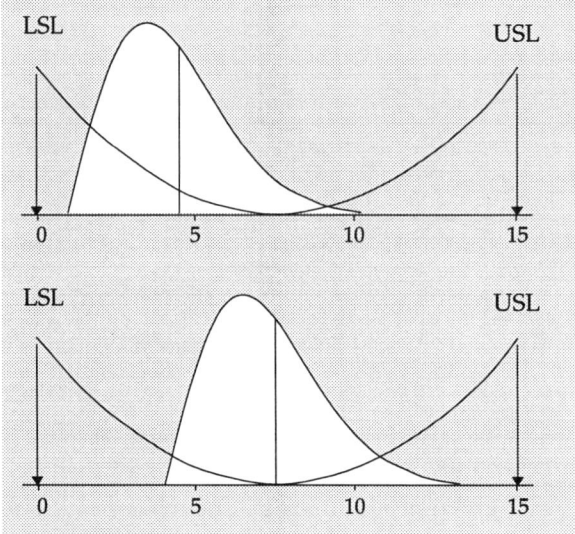

Figure 9.4: The Ball-Joint Socket Process Before and After Getting On-Target

Example 9.4: The Shaft Process

In Chapter Eight we assumed that the process for producing shafts was predictable and capable. The target value for the shaft diameter was 500 mils. Say the cost of scrapping a shaft is $1.00, and that the scrap point would be the lower specification limit of 495 mils. This gives:

$$K = \frac{C_{scrap}}{[X_{scrap} - target]^2} = \frac{\$1.00}{(495 \text{ mils} - 500 \text{ mils})^2} = \frac{\$\,0.0400}{\text{mil}^2}$$

We also assumed that the process was on target with a estimated dispersion of 1.333 mils per standard deviation unit. Combining these with the value for K we get the following Average Cost-of-Use for this process:

$$ACU = K \{ [Sigma(X)]^2 + [\bar{\bar{X}} - target]^2 \}$$

$$= \frac{\$\,0.0400}{\text{mil}^2} \{ [1.333 \text{ mils}]^2 + [500 \text{ mils} - 500 \text{ mils}]^2 \}$$

$$= \$\,0.0711 \text{ per shaft}$$

At a subsequent point in Chapter Eight we assumed that the variation for the shaft process was cut in half while remaining on-target. With this reduction in $Sigma(X)$ the Average Cost-of-Use drops from $0.0711 per shaft to:

$$ACU = K \{ [Sigma(X)]^2 + [\bar{\bar{X}} - target]^2 \}$$

$$= \frac{\$\,0.0400}{\text{mil}^2} \{ [0.667 \text{ mils}]^2 + [500 \text{ mils} - 500 \text{ mils}]^2 \}$$

$$= \$\,0.0178 \text{ per shaft}$$

for a savings of $0.0533 per shaft.

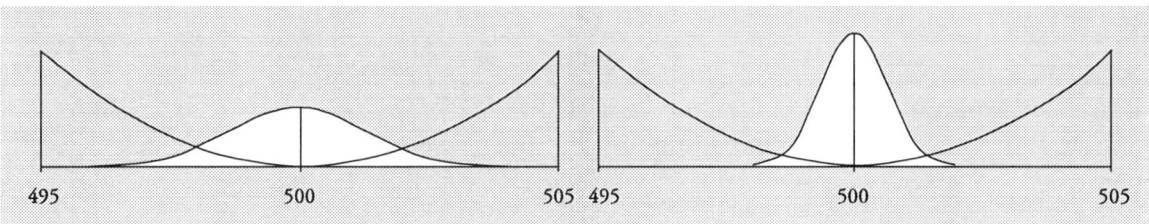

Figure 9.5: The Shaft Process Before and After the Reduction in Variation

When a product has specifications for several different characteristics, there will be an Average Cost-of-Use for each of these characteristics. These various costs may then be added together to obtain the overall Average Cost-of-Use for a given product on a per item basis.

The origin of the equation for the Average Cost-of-Use, and the justification of the formula for K will be given in the following sections. To this end we will need to begin by discussing the loss function associated with the traditional approach to quality.

9.4 The Loss Function for Conformance to Specifications

If you thought that the cost-of-using-conforming-product concept was a bit strange, you are simply a product of your upbringing. Most of us have been conditioned to think that anything within the specification limits is okay. The objective is to get the product within the specs, and anything between the uprights counts! If a value is on target, that is no better than being barely inside one of the specification limits. We could summarize this world view with the drawing in Figure 9.5—anything between the uprights counts!

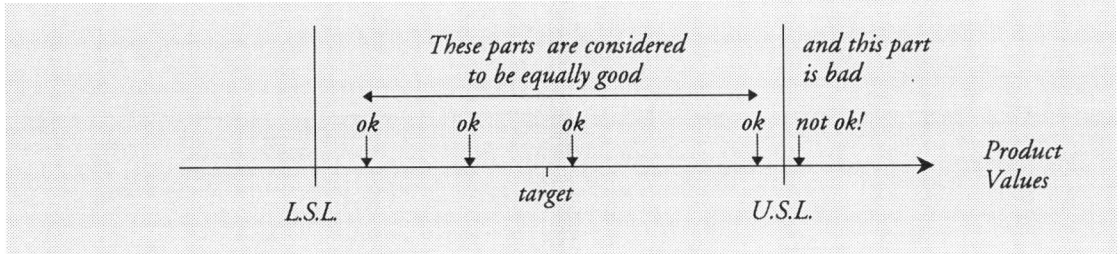

Figure 9.6: The Conformance to Specifications World View

Say, for the purposes of illustration, that the specifications in Figure 9.6 are 100 ±10, and that you are producing a stream of units that are supposed to meet these specifications. Your first unit has a value of 92 and is therefore deemed to be satisfactory. The next unit has a value of 96, and it also is passed. The third unit has a value of 101, so it is passed. The fourth unit has a value of 109, which is still within specifications, so it is passed and everything is still deemed to be satisfactory for the production process. However, when the sixth unit has a value of 111 the whole department is thrown into an uproar to find out why they are making nonconforming product! Inspectors are sent to inspect all incoming products. Engineers are assigned to project teams to work on the process. Managers consider if a recall is needed. And the workers adjust the process to increase the product values. This sudden cascade of actions will of course greatly increase the costs associated with the production of this product. (You should note that the difference between the fourth unit (109) and the fifth unit (111) was less than any of the differences between earlier successive units, yet the fourth unit was deemed to be satisfactory and the fifth unit was deemed to be unsatisfactory!)

Given the conceptualization shown in Figure 9.6 and described above, we can see that the conformance to specifications definition of quality will attach zero losses to those items that are within the specifications, and a definite, positive amount of loss to those items that fall outside the specifications. Thus, the loss function associated with the conformance to specifications definition of quality is the step function shown in Figure 9.7.

Thus, the conformance to specifications definition of quality essentially denies that there are any losses associated with conforming product. However, as we saw in the previous chapter, there are losses associated with the use of conforming product. As the product deviates from the target value, the losses will accumulate, and finally, even the specification diehards can no longer deny that these losses exist. When this happens we draw the line, call that the specification limit, and admit that the losses exist.

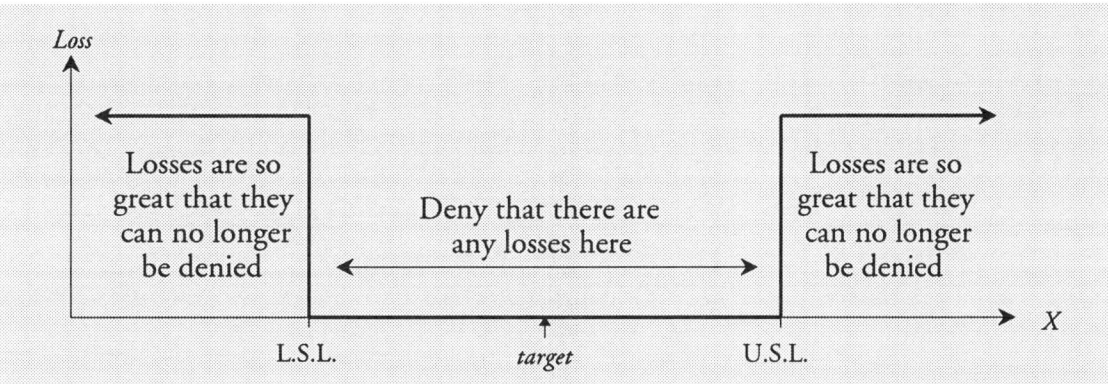

Figure 9.7: Loss Function for the Conformance to Specifications Concept of Quality

Hence, specification limits are actually *artificial* boundaries used to make *arbitrary* decisions about what product to use. They are a naive attempt to deal with the problems created by the variation of product characteristics. All product is considered to be either good or bad, and the dividing line between good stuff and bad stuff is seen to be a sharp cliff.

The world view shown in Figure 9.6 is reinforced by the cascade of actions which are initiated whenever the value for X falls outside the specification limits. Sorting, blending, rework, scrapping of product and adjusting the process will all contribute to the loss associated with nonconforming product, and this sudden shift in the way the product is treated will essentially create a step-function in any cost curve. However, we should note that this step-function is created by a reaction on the part of management, rather than by any sudden and dramatic change in the product characteristic, X. The changes from 92 to 96, from 96 to 101, and from 101 to 109 were all larger than the change from 109 to 111 which triggered the responses. According to the specification world view, 92, 96, 101 and 109 are the same, but 111 is different from 109. The shift from "operating okay" to operating "in trouble" is always seen as sudden and unexpected.

Therefore, the very nature of the conformance to specifications concept of quality fosters alternating periods of benign neglect and intense panic. During the periods of benign neglect any insights we may have gained into the process are usually lost, process improvements come unraveled, and the product quality begins to drift away from the target value once again. This is why conformance to specifications is no longer enough to remain competitive in today's world. The conformance to specifications approach does not engender the constancy of purpose required to learn from, and to continually improve, the production process. As long as the conformance to specifications is regarded as the main objective for any operation, it will be impossible to sustain any real process improvements.

Therefore, a different approach to the problem of product variation was needed.

9.5 A More Realistic Loss Function

Instead of using the step function of Figure 9.7, we could define a more realistic loss function in the following manner. If a process outcome is right on the target we will say that the loss is zero. This is because there should be no incremental cost-of-use for this outcome. However, if an outcome is not equal to the target there will be some cost associated with using this item. The greater the deviation from the target, the greater the cost of using the item. Of course, as the product value moves away from the target there will come a point where we will cut our losses, and decide to scrap or rework the item. At this point the loss function will level out, and greater deviations will no longer result in greater costs. This more realistic loss function is shown in Figure 9.8.

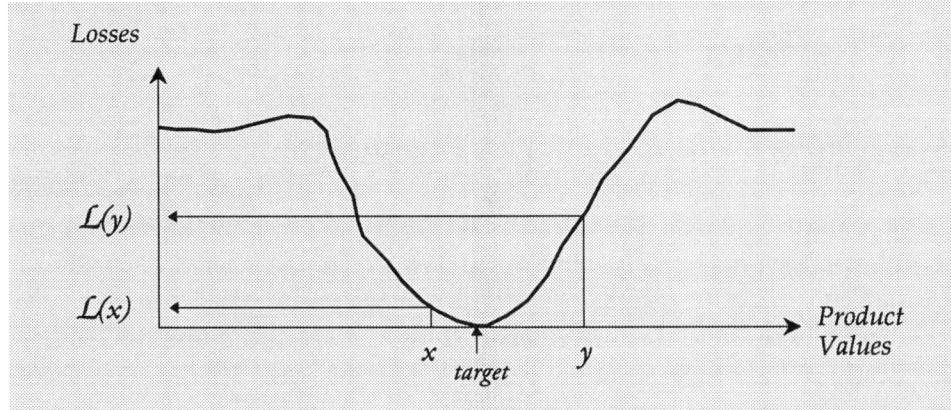

Figure 9.8: A More Realistic Loss Function

This use of a continuous curve rather than a step function is in keeping with the preponderance of all of our experience with the physical world—for most phenomena, continuous functions are more realistic than step functions.

But what are these losses? Close to the target they are the costs of using conforming product

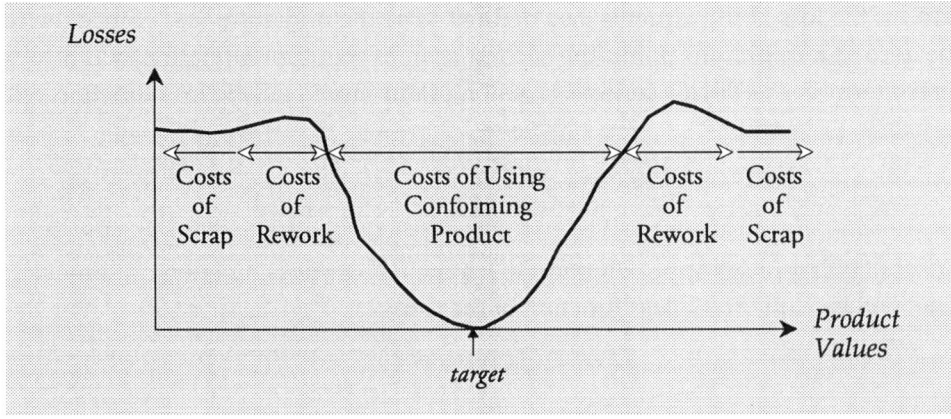

Figure 9.9: The Costs Involved in a More Realistic Loss Function

described earlier. These include difficulty in assembly, the necessity of testing and adjusting assemblies, the costs of testing and inspection that would not be needed if everything were on-target, the costs of early failures that have to be covered in warranty. Later, at the lip of the crater, these losses are the costs associated with inspection and rework of marginal units. Finally, far from the target, the losses represent the costs associated with scrapping nonconforming product.

Now how could you use this more realistic loss function? Well, if you actually had your loss function precisely defined, and if you could test a group of items to determine the product values, then you could use the more realistic loss function to evaluate the cost of using each item, and you could then average these costs-of-use to obtain an average cost-of-use for the group of items. If this group of items consisted of the whole of your process output, then this average cost-of-use would represent an opportunity available to you. It would be the potential savings that you could achieve using the techniques of continual improvement.

Fortunately, we do not have to precisely define the more realistic loss function, nor do we have to measure every item produced before we can use this concept of the average cost-of-use to define our opportunities.

9.6 Approximating a More Realistic Loss Function

With the help of a little mathematics we can avoid having to precisely define the more realistic loss function. We can instead approximate it with a fairly simple equation. Over 250 years ago Brook Taylor showed us that under some rather general conditions we may approximate a function such as the one in Figure 9.8 with a series expansion. Letting τ denote the target value, the first three terms of this series expansion are:

$$\mathcal{L}(x) \quad \approx \quad \mathcal{L}(\tau) \quad + \quad \mathcal{L}'(\tau)\,[\,(x - \tau)\,] \quad + \quad \frac{\mathcal{L}''(\tau)}{2}\,[\,(x - \tau)^2\,] \quad + \quad \cdots$$

| loss for some value x | is approx equal to | loss at target value | + | first derivative times deviation from target | + | half of second derivative times squared deviation from target | + | terms of higher order |

And this series expansion will provide, in some neighborhood of the target, a reasonable approximation to the more realistic loss function.

The loss at the target is defined to be zero—there is no cost-of-use associated with an item that falls right at the target value. This means that the first term on the right side of the approximation above will vanish. Moreover, the loss at the target is also defined to be the minimum loss. Since the first derivative will always be zero at a minimum point, the second term in our approximation will also vanish. Thus, the first non-zero term in the Taylor series expansion of our more realistic loss function will be the third term:

$$\mathcal{L}(x) \quad \approx \quad \frac{\mathcal{L}''(\tau)}{2}\,[\,(x - \tau)^2\,] \quad + \quad \cdots$$

This means that a first-order approximation to a more realistic loss function, in some neighborhood of the target value, will be a quadratic loss function of the form:

$$\mathcal{L}(x) \approx Q\mathcal{L}(x) = K\,(x - \tau)^2$$

where the constant K is a value that converts the squared deviation into dollars.

Beyond Capability Confusion

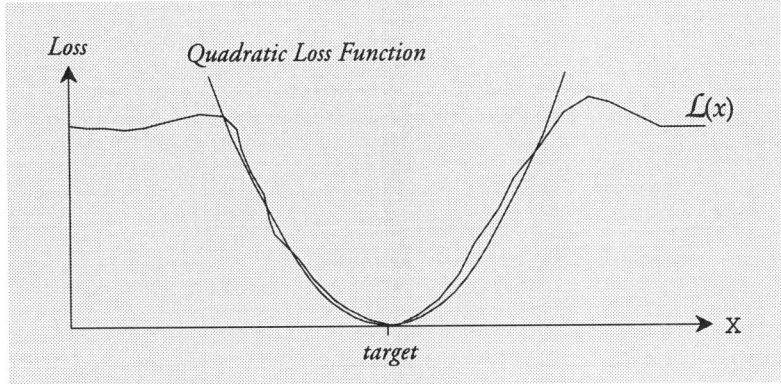

Figure 9.10: A Quadratic Approximation to a More Realistic Loss Function

It should be emphasized that the more realistic loss function, $\mathcal{L}(x)$, is not likely to be approximated by the quadratic loss function over all values for X. For example, there is probably some maximum loss that will eventually place an upper bound upon $\mathcal{L}(x)$. However, even though $\mathcal{L}(x)$ may well be more complex than the simple quadratic loss function defined above, the quadratic loss function shown in Figure 9.10 will usually provide a sufficiently good approximation in some region close to the target.

Since the cost of using conforming items is concerned with values that are close to the target, the Taylor series expansion argument used above tells us that we can, in effect, use a quadratic loss function to determine the Average Cost-of-Use for a predictable process.

9.7 The Average Cost-of-Use for a Predictable Process

If we only produced two units, and the product values for these two units were the values x and y, then the Average Cost-of-Use for these two items would be the average of the two losses:

$$\text{Average Cost-of-Use} = ACU = \frac{Q\mathcal{L}(x) + Q\mathcal{L}(y)}{2}$$

We can extend the notion above to more than two items. However, as the number of items increases

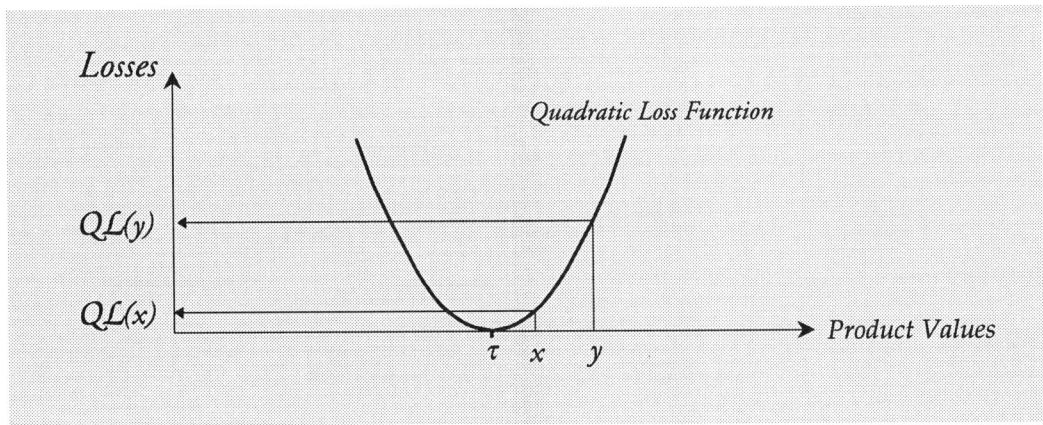

Figure 9.11: A Quadratic Loss Function and the Losses Associated with Two Units

we will find it useful to use a shortcut to finding the Average Cost-of-Use. Instead of working with the value of each item, we could represent the range of product values produced by a predictable process using some probability function, $f(x)$. And instead of computing the quadratic loss associated with each item, we could multiply the probability function by the quadratic loss function to get the double humped curve shown in Figure 9.12. When using this short-cut, the Average Cost-of-Use for the product produced by the predictable process will be equal to the area under the curve in the lower portion of Figure 9.12.

$$\text{Average Cost-of-Use} = \text{Area under curve obtained by multiplying } Q\mathcal{L}(x) \text{ and } f(x)$$

$$= \int_{\text{all } x} Q\mathcal{L}(x) f(x) \, dx$$

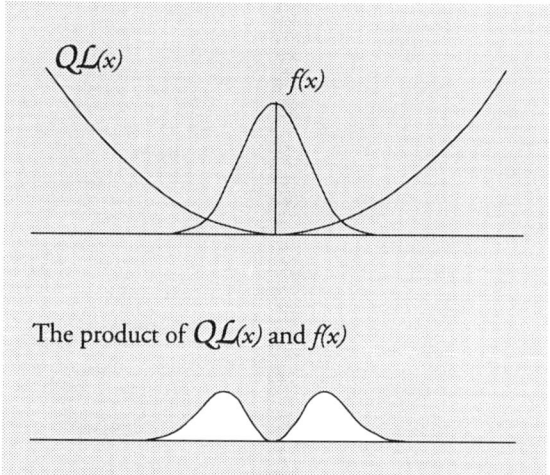

Figure 9.12: The Average Cost-of-Use is
the Area Under the Curve Obtained by Multiplying
the Quadratic Loss and the Product Distribution

Using the representation in Figure 9.12, we can see what would happen to the Average Cost-of-Use if the process were to drift off target. The greater losses that would be attached to the different product values would result in a dramatic increase in the Average Loss, as shown in Figure 9.13. Likewise, if the process dispersion were to increase, then the Average Cost-of-Use would also increase, as shown in Figure 9.14. While Figures 9.13 and 9.14 suggest that the Average Cost-of-Use is sensitive to process location and process dispersion, the exact relationship is more easily seen if we complete the integral given earlier.

$$\text{Average Cost-of-Use} = \int_{\text{all } x} K(x-\tau)^2 f(x) \, dx = K\,[\,\sigma^2 + (\mu-\tau)^2\,]$$

where

σ^2 is the *square of the standard deviation* (or *variance*) of the distribution of X

and

$(\mu - \tau)^2$ is the *square of the bias* of the distribution of X

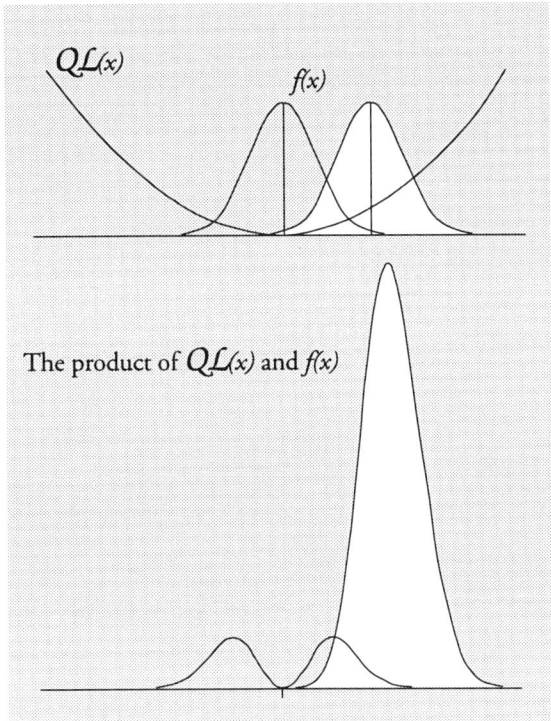
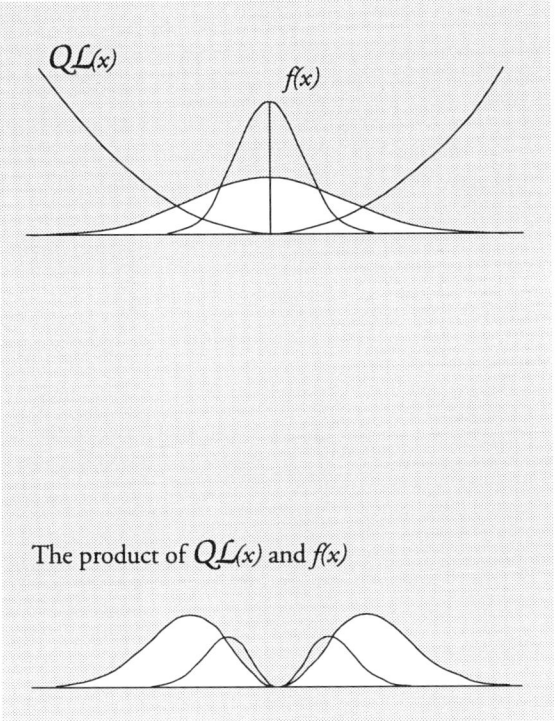

Figure 9.13: The Average Cost-of-Use Will Increase When the Process is Off Target,

Figure 9.14: The Average Cost-of-Use Will Increase with Greater Process Dispersion

The result on the previous page holds regardless of what probability function, $f(x)$, that you might use. Thus, for a predictable process, the Average Cost-of-Use may be expressed in terms of characteristics of the distribution of process outcomes.

9.8 World-Class Quality

Since the Average Cost-of-Use is an actual cost of production, its reduction will be a desirable objective, and the equation above suggests a particular strategy for doing this. The Average Loss depends upon the (1) process variance and (2) the square of the process bias. Therefore,

*the only way
that you can minimize the Average Cost-of-Use
is to operate your process
"on-target with minimum variance."*

In terms of an existing process, you will operate on-target only when the process aim is properly set. Thus you will need some way of knowing at what level your process is operating, and process behavior charts are the simplest way of tracking this.

And you will operate with minimum variation only when you operate your process predictably. (A predictable process is operating as consistently as it currently can. An unpredictable process will not be operating with minimum variance.)

Notice what we have done here. Instead of starting with some preconceived notion of quality, we have simply created a model of reality by using a more realistic loss function. Based on this more realistic loss function, and some basic mathematics, we have come up with an equation that quantifies the costs associated with variation. Given this understanding of the costs of variation, the way to reduce them becomes clear, and we arrive at a new definition of quality: On-Target With Minimum Variance.

This origin of On-Target With Minimum Variance sets it apart from all other definitions of quality:

- On-Target With Minimum Variance is not wishes and hopes.
- On-Target With Minimum Variance is not somebody's snake oil.
- On-Target With Minimum Variance is not the management fad of the month.
- On-Target With Minimum Variance is the natural consequence of a more realistic loss function and some basic mathematics.

Example 9.5:   The Cover of This Book

The diagram on the cover of this book characterizes a process that was operating in the Ideal State. It was also operating On-Target with Minimum Variance. This is Process A. However, as a part began to wear the process began to drift until, on September 26, 1980, it was operating like Process B. The process had drifted 2.7 sigma units above the target value. However, since Process A was a "six-sigma" process, and then some, Process B still had a C_{pk} value of 1.35 after the drift. According to virtually all of the specification-based definitions of quality, Process B was still a quality process. Yet, the Average Cost-of-Use for Process B was dramatically greater than that for Process A:

For Process A:

$$ACU \;=\; K \left\{ \; [\,Sigma(X)\,]^2 \;+\; [\,\bar{\bar{X}} - \text{target}\,]^2 \; \right\}$$

$$=\; \frac{2000\,¥}{\text{mm}^2} \left\{ \; [\,0.0148\text{ mm}\,]^2 \;+\; [\,15.899\text{ mm} - 15.9\text{ mm}\,]^2 \; \right\} \;=\; ¥\,0.44 \text{ per piece}$$

For Process B:

$$ACU \;=\; K \left\{ \; [\,Sigma(X)\,]^2 \;+\; [\,\bar{\bar{X}} - \text{target}\,]^2 \; \right\}$$

$$=\; \frac{2000\,¥}{\text{mm}^2} \left\{ \; [\,0.0148\text{ mm}\,]^2 \;+\; [\,15.94\text{ mm} - 15.9\text{ mm}\,]^2 \; \right\} \;=\; ¥\,3.65 \text{ per piece}$$

And, as noted on the cover, the second of these two costs is 8.29 times greater than the first. With a production volume of 17,000 pieces per day, the difference in these two Average Costs-of-Use amounted to ¥123,930 each day. While both Process A and Process B would be acceptable under any of the specification-based definitons of quality, they did not cost the producer the same amount. Perhaps this is why the producer immediately replaced the worn part, got back on target, and continued to operate with minimum variance. For more information about this example, see the Tokai Rika Cigar Lighter Socket example in Chapter Seven of *Understanding Statistical Process Control, Second Edition*, by Wheeler and Chambers.

As can be seen above, the consequences of On-Target With Minimum Variance are different from those of the other common definitions of quality. In particular, no other definition of quality will allow you to minimize the Average Cost-of-Use.

Beyond Capability Confusion

- Conformance to Requirements will not allow you to minimize the Average Cost-of-Use.
- Zero Defects will not allow you to minimize the Average Cost-of-Use.
- The Cost of Quality will not allow you to minimize the Average Cost-of-Use
- The Six-Sigma Program will not allow you to minimize the Average Cost-of-Use.
- Capability Targets will not allow you to minimize the Average Cost-of-Use.
- No specification based nostrum will allow you to minimize the Average Cost-of-Use.

The reason that these alternative definitions of quality will not allow you to minimize the cost-of-use is that they all, by means of their implicit loss function, simply deny that there are any costs of using conforming product. Yet the Average Cost-of-Use is a real cost, as was illustrated in the previous chapter. The inability to minimize this cost will limit the benefits you can receive from the traditional, specification-based notions of quality.

While the mathematical argument, and the use of the quadratic loss function, was pioneered by Karl Friedrich Gauss in 1805, the application to the problem of production was made by Genichi Taguchi in 1960. World-class quality has been defined by On-Target With Minimum Variance *for nearly forty years!* The sooner you wake up to this fact of life, the sooner you can begin to compete.

The mathematics are beyond debate. The costs are real. There is no question that On-Target With Minimum Variance is a better way to increase quality while lowering costs. The only question is whether or not you will respond by adopting the new definition of quality. If you do, then you will have to begin the journey of continual improvement. If not, then you will suffer the consequences.

The failure to operate a process On-Target With Minimum Variance will inevitably result in dramatic increases in the Average Cost-of-Use for your product. Such losses are always unnecessary. The choice is yours.

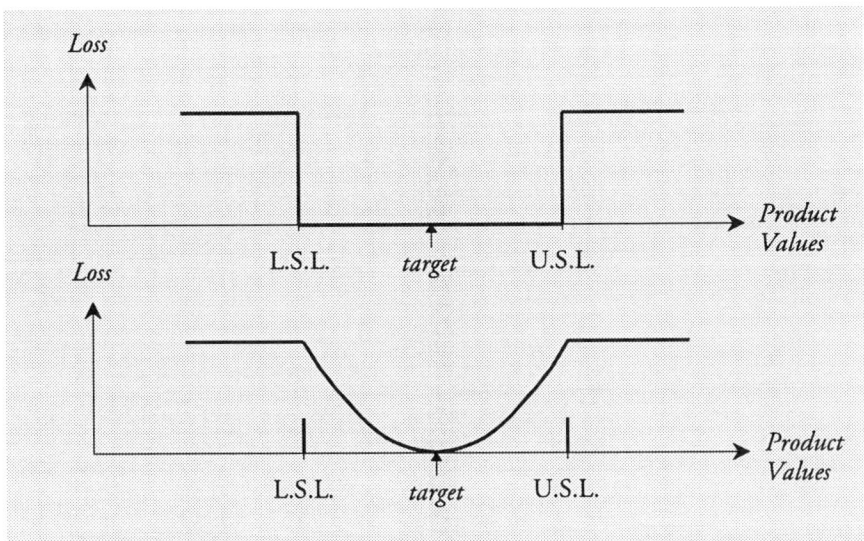

Figure 9.15: Which Loss Function is More Realistic?

9.9 Computing the Average Cost-of-Use for Your Process

Based upon the equation for the Average Cost-of-Use found in the preceding section we can compute the Average Loss for a predictable process using the information provided by the process behavior chart:

$$\text{Average Cost-of-Use} = K \left[(Sigma(X))^2 + (\bar{\bar{X}} - \tau)^2 \right]$$

where $\bar{\bar{X}}$ is the Grand Average from the average and range chart, or the average from the XmR chart;
$Sigma(X)$ is a within-subgroup measure of dispersion as defined in Section 4.4;
and τ denotes the target value for this product characteristic.

For an unpredictable process you can compute an Average Cost-of-Use value to use to describe the past. To do this you would use some descriptive measure of location in place of the Grand Average and some descriptive measure of dispersion in place of $Sigma(X)$ in the formula above. See Examples 9.1 and 9.8 for illustrations of how to use such cost values.

The value for K in the equation above is a conversion factor: it converts the expression in brackets from measurement units squared into dollars. In order to find a value for K you will need to be able to connect a definite cost with a specific deviation from target.

A simple way to do this is to obtain the cost of scrapping an item, and then connect this cost with the product value which will result in scrap (this might be one of the specification limits).

Once a cost, C_{scrap}, is connected with a specific value for X, say X_{scrap}, then:

$$C_{scrap} = K (X_{scrap} - \tau)^2$$

which defines a point on the quadratic loss function. By symmetry, a second point on the quadratic loss function is also known. Since the loss is defined to be zero at the target, we know three points which is sufficient to fully define a quadratic loss function for our process.

Solving the equation above for K gives:

$$K = \frac{C_{scrap}}{(x_{scrap} - \tau)^2}$$

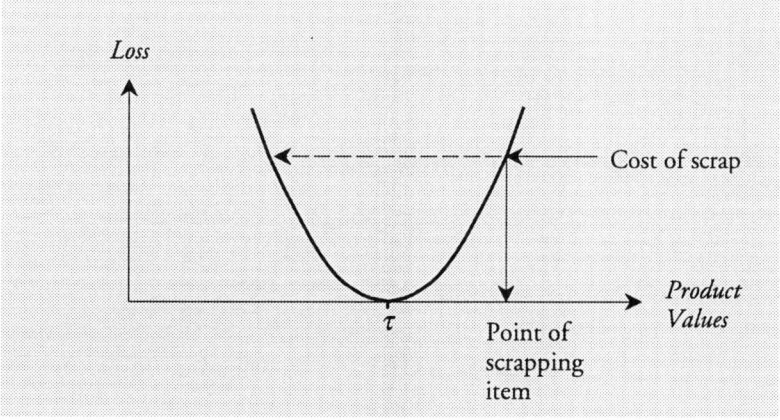

Figure 9.16: Finding a Value for the Constant K

Beyond Capability Confusion

It should be noted that using the cost of scrap in this equation is simply a convenient way of obtaining a value for K. While this will suffice to allow you to compute an Average Cost-of-Use that will characterize many of the costs of using conforming items, it will probably fail to capture all of them. Thus, the Average Cost-of-Use:

$$\text{Average Cost-of-Use in dollars} = K \left[(Sigma(X))^2 + (\bar{\bar{X}} - \tau)^2 \right]$$

should be interpreted as a lower bound on the Average Cost-of-Use for your process.

Occasionally the quantity in brackets will be computed and used without converting it into dollars. The name for this quantity is the Mean Square Deviation About Target:

$$MSD(\tau) = \left[(Sigma(X))^2 + (\bar{\bar{X}} - \tau)^2 \right]$$

The Mean Square Deviation About Target characterizes how a stable process is performing relative to On-Target with Minimum Variance. While this measure can be used to compare a process with itself at different points in time, you will get more attention, and more interest, if you use the value of K to convert the $MSD(\tau)$ value into dollars.

9.10 But What About a Loss Function That Is Not Symmetric?

In many situations the two specification limits represent different types of problems. On one side you may have to scrap a item, while on the other side you can rework the item. And then there are the cases where the loss function is one-sided simply because there is only one specification limit. When these situations exist the loss function will not be symmetric. Fortunately, because of the generality of the argument, we can use a two-part quadratic loss function to approximate a nonsymmetric loss function.

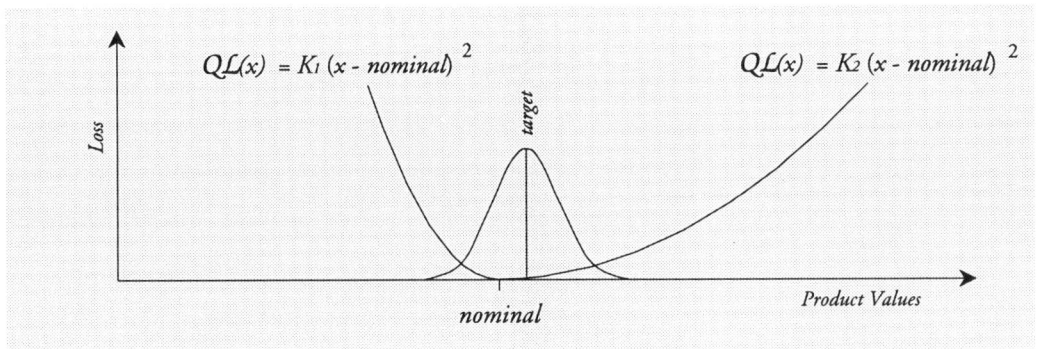

Figure 9.17: Two Quadratic Functions with Minimum at the Nominal Value

With a nonsymmetric loss function we must make a distinction between the *nominal* dimension and the *target* value. The nominal dimension would be that point where the losses are defined to be zero. With a nonsymmetric loss function you will want to have a product distribution that is centered slightly to one side of the nominal in order to minimize the Average Cost-of-Use. Hence, your target value will be slightly different from the nominal value. (Of course, when the loss function is symmetric, the target value will be the nominal value, which is why the two were not distinguished in the early part of this chapter.)

Thus, in order to obtain a nonsymmetric loss function you will have to define two quadratic loss functions, one on each side of the nominal value, using two different values of K to reflect the different types of problems.

The value for K_1 would be found using C_1, the cost associated with an item being nonconforming at the lower specification limit, LSL:

$$K_1 = \frac{C_1}{(\text{LSL} - \text{nominal})^2}$$

While the value for K_2 would be found using C_2, the cost associated with an item being nonconforming at the upper specification limit, USL:

$$K_2 = \frac{C_2}{(\text{USL} - \text{nominal})^2}$$

The resulting approximation to the loss function would then look like the curve in Figure 9.18. (In the case of a one-sided specification the K factor for the side without a specification would be set to zero.)

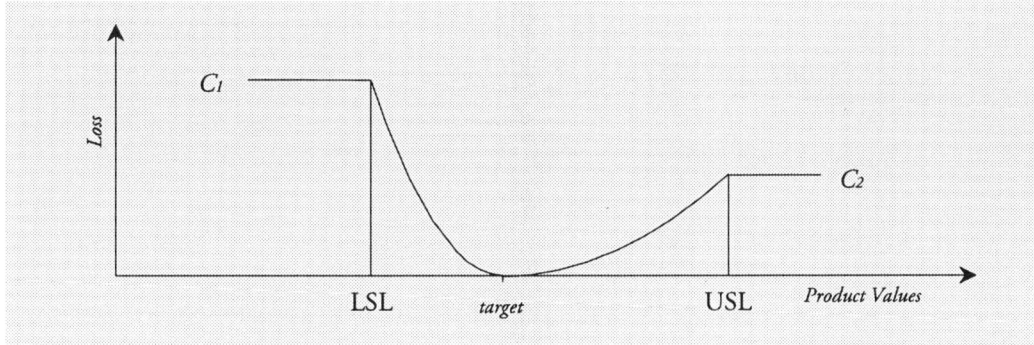

Figure 9.18: A Two-Part Quadratic Loss Function

To use a two-part quadratic loss function to compute an Average Cost-of-Use value you will need to proceed as follows:

1. You will need values for K_1 and K_2.
2. You will need a characterization of process location, such as the Grand Average, $\bar{\bar{X}}$, and a characterization of process dispersion, such as $\text{Sigma}(X) = \bar{R}/d_2$.
3. You will need to compute the Standardized Bias, B:

$$B = \frac{(\bar{\bar{X}} - \text{nominal})}{\text{Sigma}(X)}$$

4. If B is less than – 3.0, then an Average Cost-of-Use value may be obtained from the formula:

$$ACU = K_1 \, \text{Sigma}(X)^2 \, [1 + B^2]$$

5. If B is greater than + 3.0, then an Average Cost-of-Use value may be obtained from the formula:

$$ACU = K_2 \, \text{Sigma}(X)^2 \, [1 + B^2]$$

6. And if B is between – 3.0 and + 3.0, then an Average Cost-of-Use value may be obtained from:

$$ACU = K_1 \, \text{Sigma}(X)^2 \, [SAL]$$

where SAL is the Standardized Average Loss value from Table 9.2 .

The Standardized Average Loss depends upon the Standardized Bias, B, and the ratio of K_2 to K_1. Table 9.2 is set up for the formula above, where the value for K_1 is used to compute an Average Cost-of-Use.

When $K_2 = 0$ you will have a one-sided, lower specification limit. When this happens you may use the last column of Table 9.2 with the formulas above. When $K_1 = 0$ you will have a one-sided, upper specification limit. When K_1 is zero you may use the first column of Table 9.2, but the third formula above will have to be replaced by:

$$ACU = K_2 \, Sigma(X)^2 \, [\, SAL \,]$$

(Thus the first column of Table 9.2 depends on multiplication by K_2, while the remainder of the table is set up for multiplication by K_1.) The remaining columns of Table 9.2 are for two-sided specifications. When the ratio of K_2 to K_1 is 1.00 you will have a symmetric loss function.

The Standardized Average Loss values shown in boldface italics in Table 9.2 identify the approximate minimum SAL value for that loss function. The corresponding value for B defines the optimum location for the target value relative to the nominal value of 0.0. For example, when K_2 is 20 times greater than K_1, the minimum value for SAL will be near 3.349, which corresponds to a B value of −1.2, and so to achieve minimum Average Cost-of-Use you should set the target value for this process at 1.2 standard deviations below the nominal value.

A brief explanation of the origin of the values in Table 9.2 is given in the Appendix.

9 / The Average Cost-of-Use

Table 9.2 : Standardized Average Losses

B	$K_1 = 0$ *	10000	1000	100	20	10	5	4	3	2	1.50	1.00
	K_2/K_1 values											
−3.0	0.00022	**12.20**	10.22	10.00	10.00	10.00	10.00	10.00	10.00	10.00	10.00	10.00
−2.9	0.00031	12.51	9.720	9.441	9.416	9.413	9.411	9.411	9.411	9.410	9.410	9.410
−2.8	0.00044	13.26	9.282	8.884	8.848	8.844	8.842	8.841	8.841	8.840	8.840	8.840
−2.7	0.00062	14.51	8.911	8.352	8.302	8.296	8.292	8.292	8.291	8.291	8.290	8.290
−2.6	0.00087	16.45	8.628	7.864	7.777	7.768	7.763	7.763	7.762	7.761	7.760	7.760
−2.5	0.00122	19.43	8.467	7.371	7.273	7.261	7.255	7.254	7.252	7.251	7.251	7.250
−2.4	0.00168	23.59	**8.442**	6.927	6.792	6.775	6.767	6.765	6.763	6.762	6.761	6.760
−2.3	0.00232	29.50	8.609	6.520	6.334	6.311	6.299	6.297	6.295	6.292	6.291	6.290
−2.2	0.00317	37.49	9.002	6.153	5.900	5.868	5.853	5.849	5.846	5.843	5.842	5.840
−2.1	0.00430	48.36	9.701	5.835	5.492	5.449	5.427	5.423	5.419	5.414	5.412	5.410
−2.0	0.00578	62.84	10.78	5.573	5.110	5.052	5.023	5.017	5.012	5.006	5.003	5.000
−1.9	0.00773	81.87	12.33	5.375	4.757	4.680	4.641	4.633	4.625	4.618	4.614	4.610
−1.8	0.0102	106.7	14.48	5.255	4.435	4.332	4.281	4.271	4.260	4.250	4.245	4.240
−1.7	0.0135	138.8	17.37	**5.226**	4.146	4.011	3.944	3.930	3.917	3.903	3.897	3.890
−1.6	0.0176	179.8	21.17	5.305	3.895	3.719	3.631	3.613	3.595	3.578	3.569	3.560
−1.5	0.0228	230.7	25.98	5.502	3.682	3.455	3.341	3.318	3.296	3.273	3.261	3.250
−1.4	0.0293	296.0	32.23	5.861	3.517	3.224	3.077	3.048	3.019	2.989	2.975	2.960
−1.3	0.0377	379.3	40.31	6.418	3.406	3.029	2.841	2.803	2.765	2.728	2.709	2.690
−1.2	0.0478	480.6	50.22	7.175	**3.349**	2.870	2.631	2.583	2.536	2.488	2.464	2.440
−1.1	0.0601	603.3	32.26	8.161	3.352	2.751	2.450	2.390	2.330	2.270	2.240	2.210
−1.0	0.0754	755.9	77.32	9.465	3.433	2.679	2.302	2.226	2.151	2.075	2.038	2.000
−0.9	0.0938	939.3	95.47	11.09	3.591	**2.654**	2.185	2.091	1.998	1.904	1.857	1.810
−0.8	0.116	1158	117.1	13.09	3.837	2.681	2.102	1.987	1.871	1.756	1.698	1.640
−0.7	0.142	1421	143.3	15.55	41.88	2.768	2.058	1.916	1.774	1.632	1.561	1.490
−0.6	0.173	1732	174.3	18.50	4.649	2.918	**2.053**	1.879	1.706	1.533	1.447	1.360
−0.5	0.210	2097	210.7	22.00	5.233	3.137	2.089	**1.879**	1.669	1.460	1.355	1.250
−0.4	0.252	2525	253.3	26.15	5.956	3.432	2.170	1.917	**1.655**	1.412	1.286	1.160
−0.3	0.302	3022	302.9	31.00	6.831	3.809	2.299	1.996	1.694	**1.392**	1.241	1.090
−0.2	0.359	3595	360.1	36.62	7.869	4.275	2.478	2.118	1.759	1.399	**1.220**	1.040
−0.1	0.425	4252	425.7	43.10	9.087	4.836	2.710	2.285	1.860	1.435	1.223	1.010
0.0	0.500	5001	500.5	50.50	10.50	5.500	3.000	2.500	2.000	1.500	1.250	**1.000**
0.1	0.585	5849	585.3	58.91	12.12	6.274	3.350	2.765	2.180	1.595	1.302	1.010
0.2	0.681	6806	680.9	68.42	13.97	7.165	3.762	3.082	2.401	1.721	1.380	1.040
0.3	0.788	7879	788.2	79.09	16.06	8.181	4.241	3.454	2.666	1.878	1.484	1.090
0.4	0.908	9076	907.8	91.01	18.40	9.328	4.790	3.883	2.975	2.068	1.614	1.160
0.5	1.040	10404	1041	104.2	21.02	10.61	5.411	4.371	3.331	2.290	1.770	1.250
0.6	1.187	11869	1187	118.9	23.91	12.04	6.107	4.921	3.734	2.547	1.953	1.360
0.7	1.348	13480	1348	134.9	27.10	13.62	6.882	5.534	4.186	2.838	2.164	1.490
0.8	1.524	15244	1524	152.6	30.60	15.36	7.738	6.213	4.689	3.164	2.402	1.640
0.9	1.716	17163	1716	171.7	34.42	17.26	8.675	6.959	5.242	3.526	2.668	1.810
1.0	1.925	19246	1925	192.5	38.57	19.32	9.698	7.774	5.849	3.925	2.962	2.000
1.1	2.150	21499	2150	215.0	43.06	21.56	10.81	8.660	6.510	4.360	3.285	2.210
1.2	2.392	23922	2392	239.3	47.89	23.97	12.01	9.617	7.224	4.832	3.636	2.440
1.3	2.652	26523	2652	265.3	53.08	26.56	13.30	10.65	7.995	5.342	4.016	2.690
1.4	2.931	29307	2931	293.1	58.64	29.34	14.68	11.75	8.821	5.891	4.425	2.960
1.5	3.227	32273	3227	322.7	64.57	32.30	16.16	12.93	9.705	6.477	4.864	3.250
1.6	3.543	35426	3543	354.3	70.87	35.44	17.73	14.19	10.65	7.103	5.331	3.560
1.7	3.876	38764	3876	387.7	77.54	38.78	19.40	15.52	11.64	7.766	5.828	3.890
1.8	4.230	42298	4230	423.0	84.61	42.31	21.16	16.93	12.70	8.470	6.355	4.240
1.9	4.602	46022	4602	460.2	92.05	46.03	23.02	18.42	13.81	9.212	6.911	4.610
2.0	4.994	49940	4994	499.4	99.89	49.95	24.98	19.98	14.99	9.994	7.497	5.000
2.1	5.406	54057	5406	540.6	108.1	54.06	27.03	21.63	16.22	10.82	8.113	5.410
2.2	5.837	58371	5837	283.7	116.7	58.37	29.19	23.35	17.51	11.68	8.759	5.840
2.3	6.288	62875	6287	628.8	125.8	62.88	31.44	25.15	18.86	12.58	9.434	6.290
2.4	6.759	67586	6759	675.9	135.2	67.59	33.79	27.04	20.28	13.52	10.14	6.760
2.5	7.249	72486	7249	724.9	145.0	72.49	36.24	29.00	21.75	14.50	10.87	7.250
2.6	7.759	77592	7759	775.9	155.2	77.59	38.80	31.04	23.28	15.52	11.64	7.760
2.7	8.289	82892	8289	828.9	165.8	82.89	41.45	33.16	24.87	16.58	12.43	8.290
2.8	8.840	88400	8840	884.0	176.8	88.40	44.20	35.36	26.52	17.68	13.26	8.840
2.9	9.410	94097	9410	941.0	188.2	94.10	47.05	37.64	28.23	18.82	14.11	9.410
3.0	10.00	99998	10000	1000	200.0	100.0	50.00	40.00	30.00	20.00	15.00	10.00

Entries in the table are the values of the Standardized Average Loss = $SAL = [\, p + (1-p) K_2/K_1 \,] [\, 1 + B^2 \,]$.
The Average Cost-of-Use may be computed using
$$ACU = K_1\, Sigma(X)^2\, [\,SAL\,].$$

* When $K_1 = 0$ the entries in the table are: $SAL = (1-p)[\, 1 + B^2 \,]$ and for this column
the Average Cost-of-Use should be computed using $ACU = K_2\, Sigma(X)^2\, [\,SAL\,]$.

Table 9.2 : Standardized Average Losses

B	K_2/K_1 values 1.00	0.67	0.50	0.33	0.25	0.20	0.10	0.05	0.01	0.001	0.0001	$K_2 = 0$ *
−3.0	10.00	10.00	10.00	10.00	10.00	10.00	10.00	10.00	10.00	10.00	10.00	10.00
−2.9	9.410	9.410	9.410	9.410	9.410	9.410	9.410	9.410	9.410	9.410	9.410	9.410
−2.8	8.840	8.840	8.840	8.840	8.840	8.840	8.840	8.840	8.840	8.840	8.840	8.840
−2.7	8.290	8.290	8.290	8.290	8.290	8.290	8.290	8.290	8.290	8.290	8.290	8.290
−2.6	7.760	7.760	7.760	7.759	7.759	7.759	7.759	7.759	7.759	7.759	7.759	7.759
−2.5	7.250	7.250	7.249	7.249	7.249	7.249	7.249	7.249	7.249	7.249	7.249	7.249
−2.4	6.760	6.760	6.759	6.759	6.759	6.759	6.759	6.759	6.759	6.759	6.759	6.759
−2.3	6.290	6.289	6.289	6.288	6.288	6.288	6.288	6.288	6.288	6.287	6.287	6.287
−2.2	5.840	5.839	5.839	5.838	5.838	5.838	5.837	5.837	5.837	5.837	5.837	5.837
−2.1	5.410	5.409	5.408	5.407	5.407	5.407	5.406	5.406	5.406	5.406	5.406	5.406
−2.0	5.000	4.998	4.997	4.996	4.996	4.995	4.995	4.994	4.994	4.994	4.994	4.994
−1.9	4.610	4.607	4.606	4.605	4.604	4.604	4.603	4.603	4.602	4.602	4.602	4.602
−1.8	4.240	4.237	4.235	4.233	4.232	4.232	4.231	4.230	4.230	4.230	4.230	4.230
−1.7	3.890	3.886	3.883	3.881	3.880	3.879	3.878	3.877	3.877	3.876	3.876	3.876
−1.6	3.560	3.554	3.551	3.548	3.547	3.546	3.544	3.543	3.542	3.542	3.542	3.542
−1.5	3.250	3.242	3.239	3.235	3.233	3.232	3.230	3.228	3.227	3.227	3.227	3.227
−1.4	2.960	2.950	2.945	2.940	2.938	2.937	2.934	2.932	2.931	2.931	2.931	2.931
−1.3	2.690	2.678	2.671	2.665	2.662	2.660	2.656	2.654	2.653	2.652	2.652	2.652
−1.2	2.440	2.424	2.416	2.408	2.404	2.402	2.397	2.395	2.393	2.392	2.392	2.392
−1.1	2.210	2.190	2.180	2.170	2.165	2.162	2.156	2.153	2.150	2.150	2.150	2.150
−1.0	2.000	1.975	1.962	1.950	1.943	1.940	1.932	1.928	1.925	1.925	1.925	1.925
−0.9	1.810	1.779	1.763	1.747	1.740	1.735	1.726	1.721	1.717	1.716	1.716	1.716
−0.8	1.640	1.602	1.582	1.563	1.553	1.548	1.536	1.530	1.526	1.524	1.524	1.524
−0.7	1.490	1.443	1.419	1.395	1.384	1.376	1.362	1.355	1.349	1.348	1.348	1.348
−0.6	1.360	1.303	1.273	1.245	1.230	1.221	1.204	1.196	1.189	1.187	1.187	1.187
−0.5	1.250	1.181	1.145	1.110	1.093	1.082	1.061	1.051	1.042	1.041	1.040	1.040
−0.4	1.160	1.077	1.034	0.992	0.971	0.958	0.933	0.920	0.910	0.908	0.908	0.908
−0.3	1.090	0.990	0.939	0.888	0.863	0.848	0.818	0.803	0.791	0.788	0.788	0.788
−0.2	1.040	0.921	0.860	0.800	0.770	0.752	0.717	0.699	0.684	0.681	0.681	0.681
−0.1	1.010	0.870	0.797	0.726	0.691	0.670	0.627	0.606	0.589	0.585	0.585	0.585
0.0	*1.000*	0.835	0.750	0.667	0.625	0.600	0.550	0.525	0.505	0.501	0.500	0.500
0.1	1.010	0.817	0.718	0.620	0.571	0.542	0.484	0.454	0.431	0.426	0.425	0.425
0.2	1.040	*0.815*	0.700	0.586	0.530	0.496	0.427	0.393	0.366	0.360	0.359	0.359
0.3	1.090	0.830	*0.696*	0.565	0.499	0.460	0.381	0.342	0.310	0.303	0.302	0.302
0.4	1.160	0.860	0.706	*0.555*	0.479	0.434	0.343	0.298	0.261	0.253	0.253	0.253
0.5	1.250	0.907	0.730	0.556	*0.470*	0.418	0.314	0.262	0.220	0.211	0.210	0.210
0.6	1.360	0.968	0.767	0.568	0.470	*0.411*	0.292	0.232	0.185	0.174	0.173	0.173
0.7	1.490	1.045	0.816	0.591	0.479	0.412	0.277	0.209	0.155	0.143	0.142	0.142
0.8	1.640	1.137	0.878	0.623	0.497	0.420	0.268	0.192	0.131	0.117	0.116	0.116
0.9	1.810	1.244	0.952	0.665	0.523	0.437	*0.265*	0.180	0.111	0.0955	0.0939	0.0938
1.0	2.000	1.365	1.038	0.716	0.557	0.460	0.268	0.172	0.0947	0.0773	0.0756	0.0754
1.1	2.210	1.501	1.135	0.776	0.598	0.490	0.275	0.168	0.0816	0.0623	0.0603	0.0601
1.2	2.440	1.651	1.244	0.844	0.646	0.526	0.287	*0.167*	0.0718	0.0502	0.0481	0.0478
1.3	2.690	1.815	1.364	0.921	0.701	0.568	0.303	0.170	0.0642	0.0403	0.0379	0.0377
1.4	2.960	1.993	1.495	1.005	0.762	0.615	0.322	0.176	0.0586	0.0322	0.0296	0.0293
1.5	3.250	2.185	1.636	1.097	0.830	0.668	0.345	0.184	0.0550	0.0260	0.0231	0.0228
1.6	3.560	2.391	1.789	1.197	0.903	0.726	0.372	0.195	0.0531	0.0212	0.0180	0.0176
1.7	3.890	2.611	1.952	1.304	0.983	0.789	0.401	0.207	*0.0523*	0.0174	0.0139	0.0135
1.8	4.240	2.844	2.125	1.419	1.068	0.856	0.433	0.222	0.0526	0.0145	0.0107	0.0102
1.9	4.610	3.091	2.309	1.540	1.158	0.928	0.468	0.238	0.0538	0.0123	0.00819	0.00773
2.0	5.000	3.352	2.503	1.669	1.254	1.005	0.505	0.255	0.0557	0.0108	0.00628	0.00579
2.1	5.410	3.626	2.707	1.804	1.356	1.085	0.545	0.275	0.0584	0.00970	0.00484	0.00430
2.2	5.840	3.914	2.922	1.947	1.462	1.171	0.587	0.295	0.0615	0.00900	0.00375	0.00317
2.3	6.290	4.215	3.146	2.096	1.574	1.260	0.631	0.317	0.0652	0.00861	0.00295	0.00232
2.4	6.760	4.530	3.381	2.252	1.691	1.353	0.678	0.340	0.0693	*0.00844*	0.00236	0.00168
2.5	7.250	4.858	3.626	2.415	1.813	1.451	0.726	0.364	0.0737	0.00847	0.00194	0.00122
2.6	7.760	5.199	3.880	2.585	1.941	1.553	0.777	0.389	0.0785	0.00863	0.00165	0.00087
2.7	8.290	5.555	4.145	2.761	2.073	1.658	0.830	0.415	0.0835	0.00891	0.00145	0.00062
2.8	8.840	5.923	4.420	2.944	2.210	1.768	0.884	0.442	0.0888	0.00928	0.00133	0.00044
2.9	9.840	6.305	4.705	3.134	2.353	1.882	0.941	0.471	0.0944	0.00972	0.00125	0.00031
3.0	10.84	6.700	5.000	3.330	2.500	2.000	1.000	0.500	0.100	0.0102	*0.00122*	0.00022

Entries in the table are the values of the Standardized Average Loss = $SAL = [\, p + (1-p) K_2/K_1 \,] [\, 1 + B^2 \,]$.
The Average Cost-of-Use may be computed using
$ACU = K_1\, Sigma(X)^2\, [SAL]$.

* When $K_2 = 0$ the formula above is still appropriate.

Example 9.6: The Shaft Process

In Example 9.4 we used a symmetric loss function for the shaft process. The process for producing shafts was assumed to be predictable and capable. The target value for the shaft diameter was 500 mils and the cost of scrapping a shaft was $1.00, while the scrap point was the lower specification limit of 495 mils.

If we now consider the shafts to be salvageable at the upper specification limit, with a cost of the additional rework to be $0.25, then we can use a nonsymmetric loss function to compute the Average Cost-of-Use. This gives:

$$K_1 = \frac{C_{scrap}}{(LSL - nominal)^2} = \frac{\$1.00}{(495 \text{ mils} - 500 \text{ mils})^2} = \frac{\$0.0400}{\text{mil}^2}$$

and

$$K_2 = \frac{C_{rework}}{(USL - nominal)^2} = \frac{\$0.25}{(505 \text{ mils} - 500 \text{ mils})^2} = \frac{\$0.0100}{\text{mil}^2}$$

so that the ratio of K_2 to K_1 is 0.25.

We also assumed that the process average was at the nominal with a estimated dispersion of 1.333 mils per standard deviation unit. This gives a value for B of

$$B = \frac{(\bar{\bar{X}} - nominal)}{Sigma(X)} = 0.00$$

Using these values we get the following Average Cost-of-Use for this process:

$$ACU = K_1 \ [\ Sigma(X)\ ]^2 \ \{\ SAL\ \} = \$0.0400 \ (1.333)^2 \ \{0.625\} = \$0.0444 \text{ per shaft}$$

Table 9.2 shows that this process will have a minimum Average Cost-of-Use if the process average is shifted to correspond to B = –0.5. This would be:

$$\bar{\bar{X}} = -0.5 \ (1.333 \text{ mils}) + 500 \text{ mils} = 499.3 \text{ mils}$$

and at this process average the Average Cost-of-Use would drop to

$$ACU = K_1 \ [\ Sigma(X)\ ]^2 \ \{\ SAL\ \} = \$0.0400 \ (1.333)^2 \ \{0.470\} = \$0.0334 \text{ per shaft}$$

At a subsequent point in Chapter Eight we assumed that the variation for the shaft process was cut in half while remaining on-target. With this reduction in $Sigma(X)$, and assuming that B = 0.0, the Average Cost-of-Use drops from $0.0444 per shaft to:

$$ACU = K_1 \ [\ Sigma(X)\ ]^2 \ \{\ SAL\ \} = \$0.0400 \ (0.667)^2 \ \{0.625\} = \$0.0111 \text{ per shaft}$$

and by centering the process at 499.7 mils this Average Cost-of-Use could be further reduced to:

$$ACU = K_1 \ [\ Sigma(X)\ ]^2 \ \{\ SAL\ \} = \$0.0400 \ (0.667)^2 \ \{0.470\} = \$0.0084 \text{ per shaft}$$

These different values for the Average Cost-of-Use can be used to evaluate the different options for operating this process.

Example 9.7:   **The Thrust Bearing Problem**

In a jet aircraft engine the bearings of the turbine shafts have to operate under a considerable load. Certain diameters on these bearings are critical. If they are too small the engine will seize up, with disastrous consequences. If they are too large, then the bearing will leak oil. Thus the ratio of K_2 to K_1 is on the order of 1 to 10,000, or 0.0001. And, as common sense would dictate, Table 9.2 shows that the minimum Average Cost-of-Use will occur when the process for producing thrust bearing diameters is centered on a target that is three standard deviations above the nominal diameter. The loss function for this situation is shown in Figure 9.19.

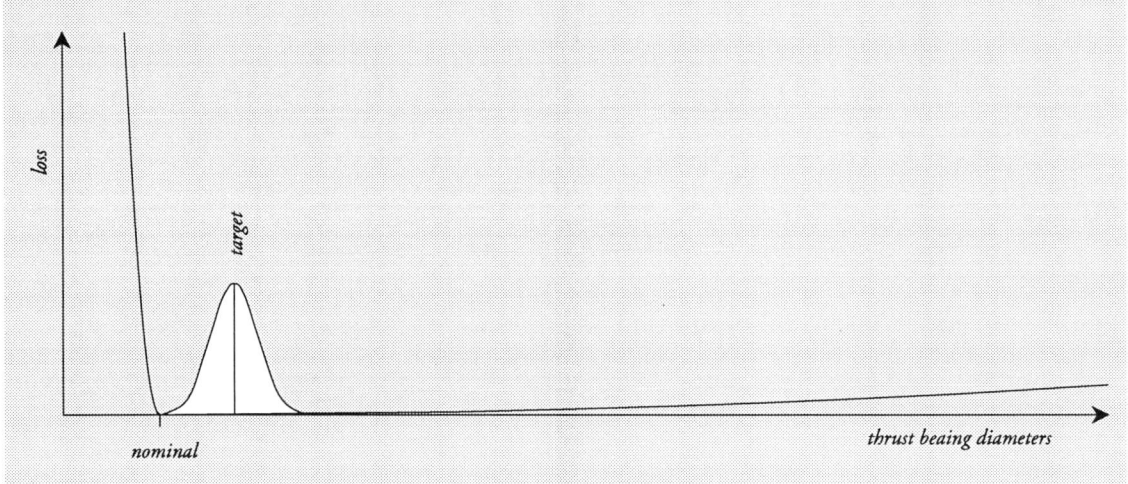

Figure 9.19: The Loss Function for Thrust Bearing Diameters

Since one of the purposes of using monetary values is to make comparisons it is appropriate to use the Average Cost-of-Use in a descriptive sense even when a process is not predictable. However, to actually achieve a predicted Average Cost-of-Use it will be necessary to operate your process predictably. For an unpredictable process, the Average Cost-of-Use may be changing, yet a value may still be computed using a global measure of dispersion (such as was used in the computation of the performance indexes). This current estimate may then be compared with a value computed using the information from a process behavior chart (which defines the hypothetical capability of the unpredictable process). The following example illustrates how this might be done.

Example 9.8:   The Batch Weight Data

In Example 5.2 we discovered that the Batch Weight Data were unpredictable over the course of 259 consecutive batches, and noted that the unpredictable weights were the visible manifestation of an unpredictable and inconsistent formulation process. The specifications used were 850 to 990, with a midpoint of 920 kilograms.

The 259 batches had an average weight of 937 kg. and a global standard deviation of $s = 61.26$ kg. Let us assume that the loss function is symmetric, and that the cost of scrapping a batch is $5,000. This yields a value for K_1 of:

$$K_1 = \frac{C_1}{(\text{LSL} - nominal)^2} = \frac{\$5000.}{(850 \text{ kg.} - 920 \text{ kg.})^2} = \$1.020 \text{ per kg.}$$

The standardized bias would be:

$$B = \frac{(\bar{\bar{X}} - nominal)}{s} = \frac{937 - 920}{61.26} = 0.3$$

So the Average Cost-of-Use for these 259 batches would be:

$$ACU = K_1 [s]^2 \{SAL\} = \$1.020 \, (61.26)^2 \{1.090\} = \$4172.35 \text{ per batch}$$

In Example 5.5 we used the median moving range of 20 kg. to define the hypothetical capability of this process. Using this value we find:

$$Sigma(X) = \frac{\widetilde{mR}}{d_4} = \frac{20 \text{ kg.}}{0.954} = 20.96$$

From this value, and assuming $B = 0.0$ is possible, we get

$$ACU = K_1 [s]^2 \{SAL\} = \$1.020 \, (20.96)^2 \{1.000\} = \$448.11 \text{ per batch}$$

for a potential savings of over $3700. per batch, which can be realized if and when they learn how to operate this process on-target with minimum variance.

9.11 Summary

On-Target with Minimum Variance is the definition of world-class quality. Once you choose a more realistic loss function, the rest is inescapable. It is mathematically certain that you will have the highest quality and the highest productivity when you operate On-Target with Minimum Variance. And the Average Cost-of-Use is the measure to use in tracking how you are doing in your efforts to operate On-Target with Minimum Variance. The details of building a continual improvement culture are beyond the scope of this book, but the need to get started down this road should be clear at this point.

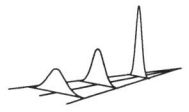

Chapter Ten

So What Do You Do Now?

10.1 The Capability Game

Your customer still wants to know if you can provide him with product that conforms to the specifications without having to resort to 100 percent screening and inspection. In Chapter One I said that the purpose of this book was to provide a self-validating way to characterize the process that is fair to both the producer and the consumer.

The process behavior chart provides a way to do just this. It can be used at production or as a way of understanding incoming inspection. When used in conjunction with a histogram of the individual measurements it is easy to characterize any product stream as belonging to one of the four possibilities. If a process is in the Ideal State, then you can count on conforming product without having to resort to inspection. If the process is in the Threshold State, then you will have a predictable level of nonconformity. Here you will need to work to change the process or to modify the specifications, and in the meantime, you will need to determine where it is most cost effective to sort out the nonconforming product. If your process is on the Brink of Chaos then you must consider the past history of 100 percent conforming product to have been a fortunate coincidence, which you cannot count on continuing in the future. The very unpredictability of your process virtually guarantees that you will get to spend some time in the State of Chaos. And you already know more about the State of Chaos than you ever wanted to know.

Capability indexes can be used as descriptive measures that characterize the Elbow Room for a process and the location of a process relative to specifications. They augment the process behavior charts. However, because they do not consider process stability, they cannot, by themselves, tell the full story. The fact that some processes are predictable, while others are unpredictable, makes the issue of capability one that cannot be addressed directly. Any discussion of capability that does not begin with a consideration of process predictability is naive, incomplete, and ultimately, misleading.

10.2 World-Class Quality

The whole issue of the capability game is focused on conformance to specifications. Conformance to specifications is the object—it is essentially the finish line—reach this point and you will have it made. Of course, all of your experience has taught you otherwise; even when you do have a brief period with 100 percent conforming product, you know things will change for the worse sooner or later. Yet this finish-line mentality is the nature of all specification-based approaches to quality—what hoops do you have to jump through in order to make the customer happy? How little can we get away with?

While capability targets, and various other programs, have raised the bar by asking for more than a minimum compliance to specifications, they still stop short of getting all that your process can deliver. They do not attempt to minimize the Average Cost-of-Use.

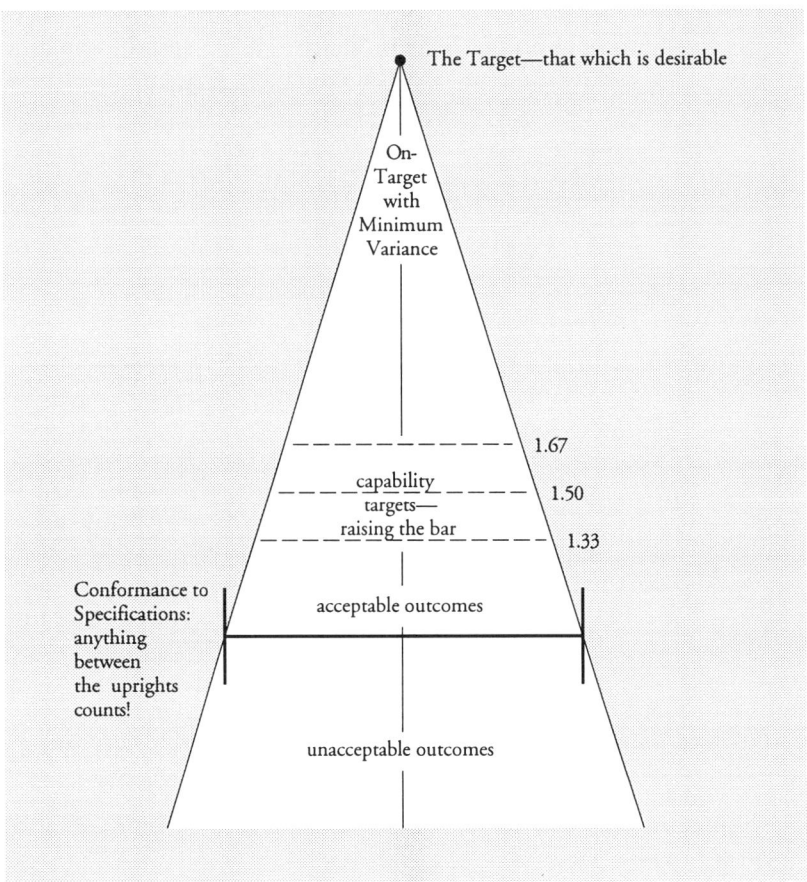

Figure 10.1: Two Approaches to Quality

Which is why there are two distinctly different approaches to quality and productivity. On the one hand you have the traditional approaches that regard quality and productivity as contrary, so it is natural to seek to do as little as possible. On the other hand you have On-Target with Minimum Variance. This does not ask for you to do more than your process is capable of doing, but it does require that you learn to operate your process predictably in order to get all that your process can do. And it delivers the highest quality with the lowest production cost. Rather than selling your process short, On-Target with Minimum Variance helps you to get the most out of your process.

10.3 "Be Ye Warmed and Filled" Is Not Enough

It is easy to set goals for others to attain. The world today is full of people and groups of people who are busy setting goals for others to meet. "Do this and you will get that" is nice, simple, and neat. Unfortunately, the problem is not with the goal, but rather with how to actually "do this." Before you will have a complete plan you will need more than a goal. A complete plan will require both a method for moving toward that goal, and a method of measuring how close you have come to the goal.

When it comes to operating On-Target with Minimum Variance, the methodology is provided by the effective use of process behavior charts in a program of continual improvement. And the Average Cost-of-Use will provide you with a scorecard measure on how close you have come to the objective of economically producing perfectly uniform product. While this book has not been intended to explain how to build an organizational culture around continual improvement, it has sought to explain the role of capability indexes and the Average Cost-of-Use in the context of continual improvement.

So there is a plan for achieving world-class quality. It does use capability indexes, but it does not use them as a club to beat the vendor over the head. The plan does use process behavior charts, but it does not use them as mere process adjustment mechanisms. The plan does seek to facilitate the conformance to specifications, but it does not see conformance to specifications as the finish line. On Target with Minimum Variance is the objective. The Average Cost-of-Use is the metric. An advantageous competitive position is the result.

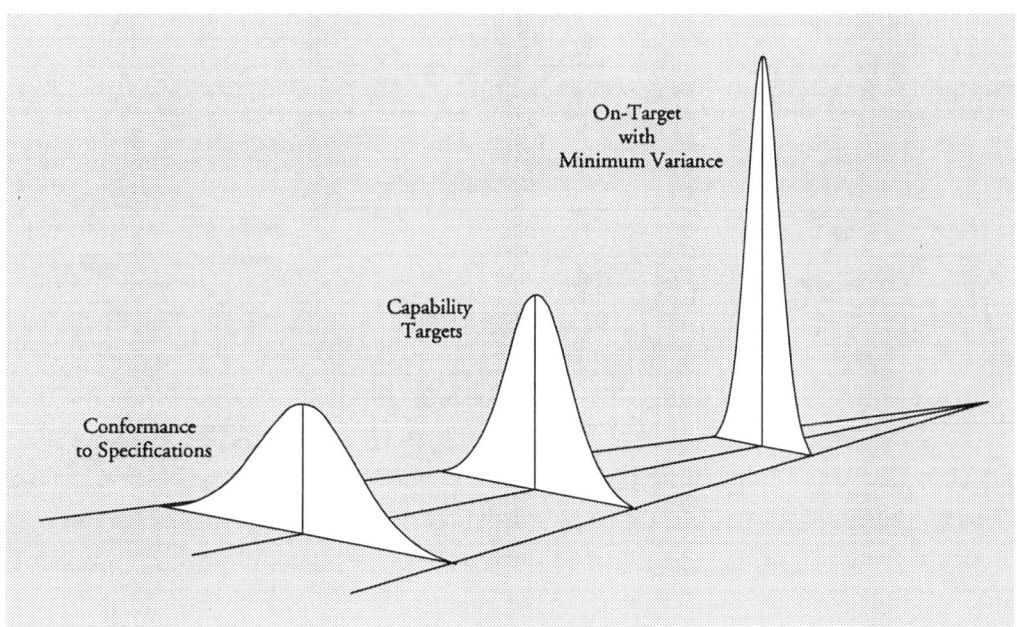

Appendix

For Further Reading...

Deming, W. Edwards, *Out of the Crisis*, Massachusetts Institute of Technology, Center for Advanced Engineering Study, Cambridge, Massachusetts, 1986.

Deming, W. Edwards, *The New Economics, Second Edition*, Massachusetts Institute of Technology, Center for Advanced Engineering Study, Cambridge, Massachusetts, 1994.

Joiner, Brian L., *Fourth Generation Management; The New Business Consciousness*, The McGraw-Hill Companies, New York, 1994.

Neave, Henry R., *The Deming Dimension*, SPC Press, Inc., Knoxville, Tennessee, 1990.

Scherkenbach, William W., *Deming's Road to Continual Improvement*, SPC Press, Inc., Knoxville, Tennessee, 1991.

Shewhart, Walter, (1931) *Economic Control of Quality of Manufactured Product*, republished by American Society for Quality, Milwaukee, Wisconsin, 1980.

Shewhart, Walter, (1939) *Statistical Method from the Viewpoint of Quality Control*, republished by Dover Publications, New York, 1986.

Wheeler, Donald J., and David S. Chambers, *Understanding Statistical Process Control, Second Edition*, SPC Press, Inc., Knoxville, Tennessee, 1992.

Wheeler, Donald J., and Sheila R. Poling, *Building Continual Improvement*, SPC Press, Inc., Knoxville, Tennessee, 1998.

Wheeler, Donald J., *Advanced Topics in Statistical Process Control*, SPC Press, Inc., Knoxville, Tennessee, 1995.

Wheeler, Donald J., *Understanding Variation*, SPC Press, Inc., Knoxville, Tennessee, 1993.

Beyond Capability Confusion

The Origin of the Values in Table 9.2

The values in Table 9.2 were found in the following manner. The Average Cost-f-Use was taken to be:

$$ACU = \int_{-\infty}^{nominal} K_1 (x-nominal)^2 f(x)\, dx + \int_{nominal}^{\infty} K_2 (x-nominal)^2 f(x)\, dx$$

where $f(x)$ is a normal probability distribution function with mean ξ and standard deviation σ.

From this it can be shown that:

$$ACU = [p K_1 + (1-p) K_2][\sigma^2 + (\mu - nominal)^2]$$

where p denotes that proportion of the completed integral represented by the first integral:

$$p = \frac{\int_{-\infty}^{nominal} (x-nominal)^2 f(x)\, dx}{\int_{-\infty}^{\infty} (x-nominal)^2 f(x)\, dx}$$

Thus, we get:

$$ACU = K_1 \sigma^2 \{p + (1-p) K_2/K_1\}[1+B^2] = K_1 \sigma^2\, SAL$$

Where B is as defined in the text. Except for the first column of Table 9.2, the values given have the form:

$$\text{Standardized Average Loss} = SAL = \{p + (1-p) K_2/K_1\}[1+B^2]$$

The values in the first column of Table 9.2 have the form:

$$\text{Standardized Average Loss} = SAL = (1-p)[1+B^2]$$

The values of p used in the table were determined by numerical integration. These proportions may be approximated by the function:

$$p \approx 1 - F(z)$$

where $F(z)$ is the cumulative standard normal distribution evaluated at the value z where z is given by:

$$z = \frac{17 B}{9}$$

Table A.1: Bias Correction Factors

The values below are those factors which convert average ranges and median ranges into appropriate measures of dispersion, namely, either $Sigma(X)$, $Sigma(\overline{X})$ or $Sigma(R)$. *

n	d_2	d_3	d_4	n	d_2	d_3	d_4
2	1.128	0.8525	0.954	21	3.778	0.7272	3.730
3	1.693	0.8884	1.588	22	3.819	0.7199	3.771
4	2.059	0.8798	1.978	23	3.858	0.7159	3.811
5	2.326	0.8641	2.257	24	3.895	0.7121	3.847
6	2.534	0.8480	2.472	25	3.931	0.7084	3.883
7	2.704	0.8332	2.645	30	4.086	0.6927	4.037
8	2.847	0.8198	2.791	35	4.213	0.6799	4.166
9	2.970	0.8078	2.915	40	4.322	0.6692	4.274
10	3.078	0.7971	3.024	45	4.415	0.6601	4.372
11	3.173	0.7873	3.121	50	4.498	0.6521	4.450
12	3.258	0.7785	3.207	55	4.572	0.6452	4.521
13	3.336	0.7704	3.285	60	4.639	0.6389	4.591
14	3.407	0.7630	3.356	65	4.699	0.6333	4.649
15	3.472	0.7562	3.422	70	4.755	0.6283	4.707
16	3.532	0.7499	3.482	75	4.806	0.6236	4.757
17	3.588	0.7441	3.538	80	4.854	0.6194	4.806
18	3.640	0.7386	3.591	85	4.898	0.6154	4.849
19	3.689	0.7335	3.640	90	4.939	0.6118	4.892
20	3.735	0.7287	3.686	100	5.015	0.6052	4.968

Given k subgroups of size n:

$Sigma(X)$ is found using either $\dfrac{\overline{R}}{d_2}$ or $\dfrac{\widetilde{R}}{d_4}$

$Sigma(\overline{X})$ is found using either $\dfrac{\overline{R}}{d_2\sqrt{n}}$ or $\dfrac{\widetilde{R}}{d_4\sqrt{n}}$

while $Sigma(R)$ is found using either $\dfrac{d_3\overline{R}}{d_2}$ or $\dfrac{d_3\widetilde{R}}{d_4}$

For the average moving range, or the median moving range, the effective subgroup size is $n = 2$.

---

* The $Sigma(\ )$ notation used in this book is a generic representation of an adjusted dispersion statistic. Some of these formulas are shown above. While these statistics might be used to estimate the dispersion parameter of some probability distribution, they are not a representation of any such parameter. They are simply a convenient way of representing any one of a collection of within-subgroup measures of dispersion.

Table A.2: Charts for Individual Values and Moving Ranges

When a *time series* has a logical subgroup size of $n = 1$ we can plot the individual values and use a two-point moving range to measure the dispersion. The average moving range or the median moving range may then be used to compute the Natural Process Limits for the individual values and the Upper Range Limit for the moving ranges according to the formulas:

$$UNPL = \bar{X} + E_2 \bar{R} \quad \text{or} \quad \bar{X} + E_5 \tilde{R}$$
$$CL = \bar{X}$$
$$LNPL = \bar{X} - E_2 \bar{R} \quad \text{or} \quad \bar{X} - E_5 \tilde{R}$$

$$URL = D_4 \bar{R} \quad \text{or} \quad D_6 \tilde{R}$$
$$CL = \bar{R} \quad \text{or} \quad \tilde{R}$$

For two-period moving ranges these scaling factors are:

$$E_2 = \frac{3}{d_2} = 2.660 \qquad E_5 = \frac{3}{d_4} = 3.145$$

$$D_4 = \left[1 + \frac{3\,d_3}{d_2}\right] = 3.268 \qquad D_6 = \frac{d_2 + 3\,d_3}{d_4} = 3.865$$

Moving ranges which fall above the upper range limit may be taken as indications of a potential break in the time series—some sudden shift which is so large that it is unlikely to have occurred by chance.

There is no advantage to using any other measure of dispersion besides a two-point moving range. Moving ranges based on any $n > 2$ will be harder to compute and less efficient than the two-point moving ranges. Mean square successive differences will be harder to compute and more prone to being inflated by extreme values while being no more efficient than the two-point moving range. Global measures of dispersion beg the question of predictability by making a strong assumption that the data are completely homogeneous (i.e. already predictable). That is why global measures of dispersion can never be properly used to compute limits for process behavior charts.

If the data do not come from a time series, i.e., if the individual values do not possess a definite and known time order, then the use of a moving range to measure the dispersion of the data is essentially arbitrary because the order of the data is arbitrary.

Moreover, if the individual values are *known* to come from different cause systems, as would be the case in a sequence of experimental runs, then the moving ranges may not be used to construct limits even when the data constitute a time series.

Table A.3: Average and Range Charts: Factors for Using the Average Range, $\bar{R}$

n	A_2	D_3	D_4	E_2
2	1.880	--	3.268	2.660
3	1.023	--	2.574	1.772
4	0.729	--	2.282	1.457
5	0.577	--	2.114	1.290
6	0.483	--	2.004	1.184
7	0.419	0.076	1.924	1.109
8	0.373	0.136	1.864	1.054
9	0.337	0.184	1.816	1.010
10	0.308	0.223	1.777	0.975
11	0.285	0.256	1.744	0.945
12	0.266	0.283	1.717	0.921
13	0.249	0.307	1.693	0.899
14	0.235	0.328	1.672	0.881
15	0.223	0.347	1.653	0.864

Given k subgroups each with n observations with Grand Average, $\bar{\bar{X}}$ and Average Range, $\bar{R}$ use the tabled constants with the formulas below to obtain

- limits for Subgroup Averages,
- limits for Subgroup Ranges, or
- Natural Process Limits for X.

$$UAL_{\bar{X}} = \bar{\bar{X}} + A_2 \bar{R} = \text{Grand Average} + A_2 \text{ times the Average Range}$$
$$CL_{\bar{X}} = \bar{\bar{X}} = \text{the Grand Average}$$
$$LAL_{\bar{X}} = \bar{\bar{X}} - A_2 \bar{R} = \text{Grand Average} - A_2 \text{ times the Average Range}$$

$$URL_R = D_4 \bar{R} = D_4 \text{ times the Average Range}$$
$$CL_R = \bar{R} = \text{the Average Range}$$
$$LRL_R = D_3 \bar{R} = D_3 \text{ times the Average Range}$$

while Natural Process Limits for individual values may be obtained from:

$$UNPL_X = \bar{\bar{X}} + E_2 \bar{R} = \text{Grand Average} + E_2 \text{ times the Average Range}$$
$$CL_X = \bar{\bar{X}} = \text{the Grand Average}$$
$$LNPL_X = \bar{\bar{X}} - E_2 \bar{R} = \text{Grand Average} - E_2 \text{ times the Average Range}$$

The formulas for these constants are:

$$A_2 = \frac{3}{d_2 \sqrt{n}} \qquad D_3 = \left[1 - \frac{3 d_3}{d_2} \right] \qquad D_4 = \left[1 + \frac{3 d_3}{d_2} \right] \qquad E_2 = \frac{3}{d_2}$$

Table A.4: Average and Range Charts: Factors for Using the Median Range, $\tilde{R}$

n	A_4	D_5	D_6	E_5
2	2.224	--	3.865	3.145
3	1.091	--	2.745	1.889
4	0.758	--	2.375	1.517
5	0.594	--	2.179	1.329
6	0.495	--	2.055	1.214
7	0.429	0.078	1.967	1.134
8	0.380	0.139	1.901	1.075
9	0.343	0.187	1.850	1.029
10	0.314	0.227	1.809	0.992
11	0.290	0.260	1.773	0.961
12	0.270	0.288	1.744	0.935
13	0.253	0.312	1.719	0.913
14	0.239	0.333	1.697	0.894
15	0.226	0.352	1.678	0.877

Given k subgroups each with n observations with Grand Average, $\bar{\bar{X}}$ and Median Range, $\tilde{R}$ use the tabled constants with the formulas below to obtain

- limits for Subgroup Averages,
- limits for Subgroup Ranges, or
- Natural Process Limits for X.

$$UAL_{\bar{X}} = \bar{\bar{X}} + A_4 \tilde{R} = \text{Grand Average} + A_4 \text{ times the Median Range}$$
$$CL_{\bar{X}} = \bar{\bar{X}} = \text{the Grand Average}$$
$$LAL_{\bar{X}} = \bar{\bar{X}} - A_4 \tilde{R} = \text{Grand Average} - A_4 \text{ times the Median Range}$$

$$URL_R = D_6 \tilde{R} = D_6 \text{ times the Median Range}$$
$$CL_R = \tilde{R} = \text{the Median Range}$$
$$LRL_R = D_5 \tilde{R} = D_5 \text{ times the Median Range}$$

while Natural Process Limits for individual values may be obtained from:

$$UNPL_X = \bar{\bar{X}} + E_5 \tilde{R} = \text{Grand Average} + E_5 \text{ times the Median Range}$$
$$CL_X = \bar{\bar{X}} = \text{the Grand Average}$$
$$LNPL_X = \bar{\bar{X}} - E_5 \tilde{R} = \text{Grand Average} - E_5 \text{ times the Median Range}$$

The formulas for these constants are:

$$A_4 = \frac{3}{d_4 \sqrt{n}} \qquad D_5 = \frac{d_2 - 3 d_3}{d_4} \qquad D_6 = \frac{d_2 + 3 d_3}{d_4} \qquad E_5 = \frac{3}{d_4}$$